Dynamic Risk Assessment

Written by experts, this book sc___ to achieve and offer an explanation on how dynamic risk assessment (DRA) is possible and how it can benefit and work within an organization.

It will provide a holistic risk management framework (cradle to grave) approach to how employers, managers and staff alike can effectively integrate and embed DRA into their business/management processes and systems to aid service delivery and risk-based decision making. The book joins up the risk assessment done in 'slow time' in the office, to real life 'quick time' actions in the field, closing this loop, and providing a feedback and debriefing approach.

- Fully supports and complements the systematic five steps to the risk assessment process.
- Multi-disciplinary dynamic risk assessment text covering fire, policing, ambulance, coastguard, lone workers and workers in the private sector.
- Contains practical examples, tips and case studies drawn from a career in the emergency services.

Stephen Asbury is Managing Director of Corporate Risk Systems Limited, a leading international consulting and training organization. Between 1998 and 2012, he was a member of the Council of IOSH; in 2010, was presented with the IOSH President's Distinguished Service Award. He has worked in a variety of senior safety and risk management roles in employment and consultancy in over 60 countries on 6 continents in a career which spans almost 30 years. This is Stephen's fourth book.

Edmund Jacobs is a Chartered Fellow of IOSH and is currently Head of Profession in the Ministry of Justice for Corporate Fire, Health and Safety. He has worked in the field of occupational health, safety and well-being for nearly 25 years and is a member of the IOSH Board of Trustees. Edmund was Chairperson for the London Health & Safety Group (2005–2008) and was awarded the status of Honorary Member. Whilst working for the City of London Police he undertook research into the application of dynamic risk assessment for his Master's Degree in Occupational Safety and Health.

Dynamic Risk Assessment

The practical guide to making
risk-based decisions with the
3-Level Risk Management Model

Stephen Asbury
and Edmund Jacobs

Routledge
Taylor & Francis Group

LONDON AND NEW YORK

First published 2014
by Routledge
2 Park Square, Milton Park, Abingdon, Oxon, OX14 4RN

and by Routledge
711 Third Avenue, New York, NY 10017

Routledge is an imprint of the Taylor & Francis Group, an informa business

British Library Cataloguing in Publication Data
A catalogue record for this book is available from the British Library

Library of Congress Cataloging-in-Publication Data
Asbury, Stephen.
   Dynamic risk assessment. – First edition
       pages cm
   Includes bibliographical references and index.
   1. Risk management. 2. Risk assessment. I. Title.
   HD61.A83 2014
   658.15′5–dc23   2013040200

ISBN: 978-0-415-85403 0 (pbk)
ISBN: 978-1-315-85872 2 (ebk)

Typeset in Sabon by
Keystroke, Station Road, Codsall, Wolverhampton

'Carry on. The paperwork looks fine.'

Contents

Figures and tables

Figures

Tables

Case studies

Foreword

It has been interesting to see how the concept of Dynamic Risk Assessment (DRA) has developed since its emergence in the early 1990s. At the time, I was the Health and Safety Manager for the London Fire Brigade and became conscious that the then new requirement for risk assessment was not always practical or possible in the emergency scenarios fire-fighters were facing. I proposed that a dynamic approach needed to be taken which took account of the nature of fire service work. Following a paper I submitted to the then Chief and Assistant Chief Fire Officers Association health and safety committee at the request of the Home Office, I ended up chairing and guiding a small group of fire officers at a three-day meeting at Moreton-in-Marsh. There, we put together the guidance which was subsequently published by the Fire Service Inspectorate.

Much of the subsequent development elsewhere appears to be misguided when considering how the concept came about. It was always intended as part of a wider, strategic approach to the managing of fire service risk, not a stand-alone. We intended that it should diminish the level of unexpectedness and unpredictability over time, as knowledge grows. It should be used only where the situation really does need a dynamic approach, and this will be infrequently and certainly not as prevalently as appears to have become the case.

Perhaps also worth noting is the difference in making risk decisions when deliberately committing people into dangerous and perhaps unknown situations, as opposed to the type of situation which arises as an emergency, often unforeseen. People cannot be prepared in isolation to face the unknowns; this can be achieved only by the management systems of the organisation being geared towards the role and environment in which they may be exposed. Equally, I do not feel that it is possible to have stand-alone training in DRA.

If you are involved in DRA, then every lesson needs to be learned from it every time it is applied, for this is the way the organisation builds a store of risk knowledge and exposure which can then be allowed for in the development of competence, provision of equipment and general support for workers who may be exposed to unknown or changing risks.

This book puts many of the misconceptions about DRA to bed. This should enable a better understanding of what is (or was) intended by the process. DRA is not a substitute for pre-planning, safe systems of work or where a pre-work risk assessment can be made. It is a means of keeping people safe when exposed to unknown and changing, dynamic situations.

John Norton-Doyle
October 2013

Endorsements

There can be no doubt that risk assessments have proved a rational and efficient way of reducing negative outcomes in the workplace. Yet for too many, they have become a bureaucratic activity rather than a matter into which one puts deep thought. 'Have you done the risk assessments?' is a standard mantra of a superior to a subordinate when any change is being introduced, however low or high risk.

Stephen Asbury and Edmund Jacobs have addressed this flabby approach head on, and restored vigour and utility to the risk assessment process. Risk assessments, they argue, are an essential and invaluable tool of the process of managing safely, but on occasions there are unexpected situations that may arise, where static risk assessments may not provide the necessary solution in isolation.

The authors draw upon and develop the concept of Dynamic Risk Assessment (DRA), developed within emergency services in the 1990s, to show how DRAs can be used more widely within rapidly changing environments to manage developing risk.

The authors are both skilled and distinguished members of the safety profession and already enjoy high reputations among their peers. This book will confirm and even enhance those reputations. It is an academic and well-thought out work that is nevertheless interesting, applicable and eminently readable, of use to the manager or supervisor as well as those in the safety profession. It will become the classic work on the subject.

Stephen Vickers PhD FCIS FCMI FCIEA
Chief Executive
Vocational Training Charitable Trust

Dynamic Risk Assessment is an overdue addition to the safety literature, as this approach to working in hazardous settings has significant mileage beyond its origins in the Fire Service. Asbury and Jacobs have succeeded in presenting the basic ideas in a very accessible manner and clearly explain their evolution within the world of safety management. One of the book's strengths is that the dynamic component is set against the broader context of strategic and predictive risk assessment, with powerful illustrations of their interdependence. Every chapter has salient case studies and importantly, these portray the underlying hazards that suddenly emerged, not just in the obvious settings such as aviation or energy production but elsewhere, such as outdoor activity centres and estate agencies. These reveal the fundamental need for ongoing risk vigilance even when the pre-task risk assessment has been completed: Dynamic Risk Assessment provides a means of achievement and is applicable in almost every workplace. The authors have sensibly realised that their readers will need advice for implementing this method and there are useful tips throughout the book which should help to transform their framework into practice.

Rhona Flin
Professor of Applied Psychology
Industrial Psychology Research Centre
University of Aberdeen

A 'must read' for every manager in every type of organization.

Dynamic risk assessment provides a risk-based methodology for undertaking operations in real time, whilst continually ensuring that actions remain relevant to the changing context in which they are taken.

It inspires a level of awareness amongst the workforce that positively influences the safety culture and individual attitudes towards safety.

This 'safe person' approach develops a raised awareness of the risks present in the working environment and the ability to assess the appropriate steps to take in order to mitigate them, for the individual, the team and for the whole organization.

David Onigbanjo
Senior Human Resources Business Partner

In some occupations, such as adventure training, emergency services, news correspondents, etc., the nature of work inevitably involves some risk taking. Unlike a static workplace, where risk is predictable and control is relatively straightforward, these workers operate in an environment which can be hostile and fast moving. Sound judgements have to be made based upon training and experience in circumstances which often fall outside the comfort zone. Dynamic risk assessment is a process used to evaluate these circumstances as they unfold and to ensure that the safest system of work available is chosen.

This book explains the process and provides guidance on how to carry out such a dynamic risk assessment.

Martyn Grant MSc CFIOSH
Former Head of Health, Safety & Environment
Thames Valley Police
Service headquarters, Oxford, UK

I have worked in Formula 1 for all of my career, and I welcome this book from Stephen and Edmund which shares good practices from other sectors and environments I'm less familiar with. It is a most interesting read and I recommend it to you if, like me, you sometimes have to make decisions in very short time.

Andy Stevenson
Sporting Director, Race Team
Sahara Force India Formula 1
Silverstone, UK

In our hospital, we have developed early warning systems to make the unusual more predictable and more easily noticed. This leads to earlier interventions and better patient outcomes. We're pleased to share some of our ideas with you through Stephen and Edmund's book about dynamic decision making. Sharing ideas amongst the professions seems a good idea to me, and we are pleased to contribute to this compendium of cases, ideas, tips and examples for anyone required to make decisions in rapidly changing situations. It is an interesting book and I enjoyed reading it. I recommend it to you if, like me, you sometimes have to make critical decisions to save and protect lives.

Dr Craig Stenhouse
Medical Director
Burton Hospitals NHS Foundation Trust
Staffordshire, UK

It's a given in the world of professional education and training that instructors *always* need to be wary of – and prepared for – the unexpected, the unanticipated, and the unwanted. This trio is always lurking in the shadows and takes great delight in appearing at any time to interrupt the transfer of knowledge, affect participant/instructor behavior, and wreak general havoc wherever teaching/learning takes place. To combat these disruptions and distractions, instructors are taught the art of 'nimbility' – the art of being able to immediately, comfortably and smoothly adjust and adapt what they're doing to meet the challenges, changes, and roadblocks thrown up by this unholy troika.

Stephen and Edmund have produced wonderful guidance on how to adapt in fast time the carefully laid plans made in advance and in slow time. As a result, they've done an excellent job of illustrating how to proactively respond to internal or external changes that will alter how a system functions unless immediately assessed and dealt with – in other words, they've clearly shown how 'nimbility' can – and does – work.

Dr David Pelton
Instructor Development Specialist
PetroSkills LLC, Houston, Texas, USA

This exciting book expertly builds on the theory and practice of dynamic risk assessment from the Fire and Rescue Service, into a practical risk based reference resource, transferrable across organisations outside of the blue light emergency services.

Public, private and charity sector organisations that operate in environments which are dynamic in nature can use the illustrative examples and case studies within the book, to help them manage risk and usefully implement new concepts, in a meaningful and sustainable way.

I would strongly recommend that you and your senior management teams read this book before you consider applying dynamic risk assessment within your organisation or before you make changes to existing risk management approaches.

Chris Steele, CMIOSH, GIFireE
Director
C. Steele Associates Limited
Fire, Health & Safety Consultancy

Dynamic risk assessment (DRA) is commonly used to describe a process of risk assessment being carried out for an activity that is developing as it is being undertaken. The application is wide ranging, from manual handling to confined-space work.

As a practitioner, I have come to appreciate the importance of applying DRAs to everyday activities; for example, use of DRAs can be adopted to address both biomechanics and the body's ability to tolerate loads, in addition to formal manual handling risk assessments.

The principles outlined in this book provide valuable technical knowledge that will be useful when candidates are studying for such courses as the NEBOSH Diploma, where DRA is covered with regard to factors to consider while developing emergency arrangements when working in confined spaces.

It is imperative to understand not only the theory but also the application of DRAs in the workplace. This book provides a detailed understanding of the theory and, more crucially, detailed guidance on how to apply DRAs. It also provides a number of excellent case studies and clearly worked examples showing best practice that the reader may apply to their work or study.

I am therefore pleased to recommend this excellent book.

Jonathan Backhouse, CMIOSH QTLS FIfL AIEMA
NEBOSH Examiner
Middlesborough, UK

In this book, the increasing importance of Dynamic Risk Assessment, following its emergence as a concept for use in the field in the 1990s, is covered in eight chapters. In essence, it is all about decision making in a dynamic environment. It is expected that recommended readership will be drawn from a wide variety of sectors, particularly those working in front line services and decision makers will also find this essential reading as an ongoing reference book.

Chapter 1 explores the relationship between dynamic risk assessment and systematic risk assessment and its application in the field. The seven chapters which follow explore:

- the reasoning behind the emergence of dynamic risk assessment as an operational control measure
- a dynamic risk assessment risk management model
- the theory of risk assessment as it moves from the board room to the field of operations
- the key areas which must be embraced to successfully convert risk assessment into a useful management tool
- the benefits of utilising a risk based approach which will inform and improve service delivery
- how the application of dynamic risk assessment can be integrated into risk assessment training
- how management can learn from organisations who have adopted dynamic risk assessment for use in their sector and its role in recent major events.

The introduction of case studies in each of the book's chapters should provide useful 'stop off' points for readers.

Not only is the book content well chosen, but also it benefits from being written by two safety professionals with much experience of working with and in the emergency services sector. It is well presented with helpful and informative tables. I particularly like the fact that key points are brought out at the end of each chapter.

In these times, when unpredictable, adverse events have a habit of occurring, advice and guidance on and understanding of dynamic risk assessment should be welcomed and this book should provide a valuable resource in such circumstances.

Malcolm Brown MIIRSM
Health and Safety Advisor
Atmospheres Air Conditioning Services Ltd

This book is a must read for every person who manages people and situations in the office or out at the sharp end of their business; dynamic risk assessment is a skill that is vital to the success of daily work activities and the safety of an organisation's prized asset, its people.

To master the dynamic risk assessment process effectively this book will open up a whole new outlook on situational management when moving out of tried and tested knowledge based zones helping to prevent human error.

Once humans move out of their comfort zone errors begin to show, therefore to factor human into safety in the form of dynamic risk assessment will help to eliminate human error within situational changing environments effectively and safely.

We all know how quickly a situation can change with the need to move to plan B, should a plan B be available. With the application and implementation of dynamic risk assessment plan B will not be required as comprehensible contingency management will be built into the dynamic risk assessment to help with the management of change as dynamic risk assessments are a live document.

Paul Benson MSc CMIOSH MIIRSM MinstRE
Operations HSE Lead
SHELL

This book is a timely intervention for enhancing the training and development of risk assessors and decision makers in all sectors.

There is a legal requirement for suitable and sufficient risk assessments to be carried out in all predictive areas of work. In spite of the best efforts of trained risk assessors undertaking predictive risk assessments, there will always be some residual risks.

The application of dynamic risk assessment principles to cover residual risks and their associated problems linked to personal injury and associated implications will enable employers to meet their legal, moral, and economic responsibilities as far as undertaking suitable and sufficient risk assessments is concerned.

This book comes highly recommended for safety professionals, senior managers and front line staff across all sectors.

Noel R. E. Thomas MSc CFIOSH
International Health & Safety Consultant

Stephen and Edmund bring together a wide collection of ideas and approaches I have used successfully in my own career, and accordingly, I am delighted to endorse this excellent book about the three levels of understanding and assessing risks, and dynamic risk assessment in particular.

When I was younger, I devoured all there was to read about 'risk' during my IRM studies. And then I duly knocked on the Chief Fire Officer's door to find out what his 'risk appetite' was to help me to understand how much risk we could i) take or ii) needed to avoid when protecting the community we served in Staffordshire. This was one of those moments in my career I will never forget; I was given very short shrift and left his office with him standing there with a 'don't be so silly' look on his face. He said 'How on earth do I know, you [the senior management team] need to come to me as you evaluate the risks [in what Stephen and Edmund call 'slow time'], give me the options to eliminate or mitigate and I will make a decision.'

Most health and safety people have a risk matrix or some software – they pump in risk numbers, and risks are evaluated. Most organizations task a mere mortal, i.e. a safety practitioner, to evaluate and suggest how risks should be managed. But prior to this, it feels like as a profession we are brain-washed into believing that risk is all negative, that it's all about down sides, worst-case scenarios of catastrophic incidents, deaths and major injuries.

In developed countries, these types of outcome do not happen very frequently. Taking a leaf out of those IRM texts, risk is far from a down side – it is full of opportunities and requires careful evaluation to understand. Contrast this with how many HSE directors promise the board they will 'de-risk the business' to prevent health and safety incidents, treating risk like disease for which a cure is urgently sought. This approach is wrong in two ways. Firstly, our commercial colleagues relish the opportunity to find solutions to clients' requirements; their 'can do' attitude means inevitably trying new things and taking risk (mitigated of course). Secondly, the law in most countries does not actually require us to eliminate all risk. Generally speaking, the duty holder must do all that is 'reasonably practicable', rather than the greater 'practicable' duty. However, how many risk assessments do you see that have a belt-and-braces approach, without carefully considering the threat posed and a balanced approach to its mitigation?

Ultimately this overly cautious treatment of risk will lead to a dilution of resources, and so other more significant risks may fail to be managed. Another unintended consequence is that those who work in the operational arm will know what goes wrong most often, and view the belt-and-braces approach as 'butt covering' or another safety zealot going overboard!

Another area where I see that improvements can be made is in the operational area. Often a surveyor will visit site and do a ten-minute walk-around, price up the job and put in a bid to tender for work. If their submission is successful, the first time the supervisor will see the site is when he and the gang arrive on day one with their tools. But surely this is OK because supervisors in most sectors are issued with their own safety manual, containing lots of general risk assessments and other safety stuff. However, what action do we expect them to take if the general risk assessment does not list the hazard posed? So the supervisor most often soldiers on and makes decisions based on his personal risk appetite, and since he was tasked to complete the work within that shift his 'can do' attitude will prevail.

Enlightened employers will have up-skilled supervisors to evaluate risk in what Stephen and Edmund call 'fast time'. This is more than just providing a site-specific risk assessment template. It is providing them with the skills to evaluate risks in the field, and the delegated authority to make decisions.

Bad decisions come back to haunt employers – Stephen and Edmund provide lots of examples in this book. Empowering employees to make the right recommendations to treat risks is fundamentally important, so that all risks are managed to an optimum point – not too much, and certainly not too little. If you were to ask me what is the one thing I would do differently, I'd tell you that the biggest bang for your buck is to up-skill supervisors, equip them to assess, manage and monitor risk in slow and fast time. This is the approach we adopted at London 2012 – we empowered and engaged this group, and the results speak for themselves – the first ever Olympic Games build which was fatality free.

Deborah Clements
Present and Past Positions
Group Health, Safety and Well-being Manager, Vodafone Group Services Limited
Senior HSE Adviser, Aggregate Industries
Head of Health and Safety Assurance, London Olympics 2012
Strategic Risk Manager, Staffordshire Fire and Rescue Service
Safety and Risk Manager, British Midland Airways

Preface

One thousand, one hundred and forty (1140).
That's the number of hits a simple search for 'dynamic risk assessment' retrieves on the website of the Health and Safety Executive (HSE) – one of the world's most prominent regulators of workplace health and safety standards.

Thirteen point one million (13.1m).
That's the number of hits the same search on Google retrieves.

And in July 2012, Mr Justice Globe, sitting in the High Court of Justice (Queen's Bench Division), agreed that 'a dynamic risk assessment was acceptable' in the welly-wanging case of Cornish Glennroy Blair-Ford and CRS Adventures Ltd (see Chapter 1).

So how has dynamic risk assessment (DRA) slipped from the relatively narrow use by 'blue light' emergency services into multiple-millions of web hits, welly-wanging case law, and mainstream health and safety risk management; and what can it do for you? That's what this book is all about.

This book is targeted at decision makers at all levels – directors, managers, supervisors, employees and health and safety advisors in the public, private and third sectors where significant risks may be encountered. Workers at all levels of the organisation must be empowered to make the right decisions in real time. They need to know how DRA can help them to achieve this, and how to effectively implement DRA as a part of their overall approach to health and safety management.

The book provides a holistic risk management framework called The 3-Level Risk Management Model™ which shows how to integrate effectively and embed DRA into business and management systems to aid service delivery and risk-based decision making. It provides this systematic framework to facilitate joining up risk assessments made in the 'slow time' environment of the office, to real-life 'fast time' actions in the field where work takes place. It closes this loop, and provides a powerful approach to feedback, debriefing and learning. It fully supports and complements the common systematic 'five steps to risk assessment' process (see Chapter 3).

Many readers may remember the name of Suzy Lamplugh – a London estate agent who was reported missing after going to an appointment with someone calling himself 'Mr Kipper' to show him a house in Fulham in 1986. She was officially declared dead, presumed murdered, in 1994 (see Chapter 6). The safety of lone workers such as estate agents can be enhanced by effective use of DRA.

Other readers may recall the quick-thinking, well-reasoned actions of pilot Chesley 'Sully' Sullenberger III, which saved the lives of all 155 passengers and crew when he successfully ditched US Airways Flight 1549 – which had been disabled by striking a flock of Canada geese during its initial climb out of LaGuardia airport – into the Hudson River off Manhattan, New York City, on 15 January 2009 (see Chapter 2). Understanding risks, and how to act dynamically saved the day. Sully was a hero (although he denies this himself), and lived to tell the tale.

Less familiar to most readers will be Max Palmer, a 10-year-old boy who drowned in May 2002 while 'plunge pooling' on a school trip in Glennridding Beck, Cumbria (see Chapter 2). The investigation revealed serious errors of judgement by the party leader in planning and leading the activity. Properly connecting the 'risk assessment' made in advance to the dynamic and changing conditions on the day might have resulted in a happier ending to the trip.

So how is this book structured?

Our introduction and explanation of DRA commences in Chapter 1 with an explanation of the associated psychology. In Chapter 1 we cover:

- the emergence and broad-brush concepts of DRA
- myths, conceptions and misconceptions
- its relationship with the legal requirements for risk assessment
- its application in the field.

Chapter 1 also introduces the first of our eighteen case studies highlighting where the informed use of DRA has mitigated losses, and also where losses have been magnified by failing to respond to escalating situations in an appropriate and timely manner.

Chapter 2 explains how DRA has moved from its origins in blue-light emergency services and other higher-risk environments to less hazardous, non-emergency environments in a host of other sectors.

Our third chapter discusses how the principles of risk assessment undertaken in slow time can be utilised to inform risk-based decision making or decision making in the field in fast time, utilising the principles of the popular 'five steps to risk assessment' approach. It explores how organisations can help their staff to be safe in dynamic, changing or unfamiliar situations by enabling and empowering them to make effective risk-based decisions.

In Chapter 4, we introduce our model for risk management, set at the heart and all levels of the organisation, which covers strategic, predictive and dynamic levels of assessment. We will show how The 3-Level Risk Management Model™ can be integrated into all types of structures for business and organisational success. The chapter also covers how the risk-assessment loop for integration and continual improvement can be achieved in practice.

Chapter 5 explores the key areas that need to be considered and accepted in order to create the right environment within and around any organisation considering utilising The 3-Level Risk Management Model™ as a risk management tool. These include:

- securing senior management support
- developing the organisation's culture
- embedding core values – how to nurture the 'right' behaviours
- employers' expectations
- the 'safe person' concept – employer and employee role.

In Chapter 6, we will explain how to use DRA to improve service delivery. We will explore the benefits of utilising a sensible and proportionate risk-based approach to achieve service delivery and effectiveness. The chapter will address the common issues of:

- risk-taking behaviour
- risk perception
- red mist (or 'rushing in').

As we head toward the conclusion of the book, Chapter 7 explains how the application of SRA (strategic risk assessment)/PRA (predictive risk assessment)/DRA can be integrated within organisations' general risk assessment training. We explain how DRA can only be applied effectively with an informedness of the complete risk assessment process. We'll explain its applicability to decision makers such as directors, managers, supervisors, workers and health and safety practitioners to aid their risk-based decision making. Finally, we will consider the implications and interfaces for shared activities and operations where DRA is not utilised in the other organisation.

Our conclusions are set out in Chapter 8, where we discuss integrating learning from your and other organisations to improve performance. We demonstrate the value of feedback, debriefing and learning as a means of providing continual improvement. We continue to analyse organisations that have applied DRA within their risk management approach and discuss societal expectations for the emergency services (i.e. that they will dash into burning buildings to save people and puppies, irrespective of the risk; like their X-BOX, TV and film heroes do), and of other hazard-facing organisations that have to deal with real-time situations while balancing the protection of staff and achieving operational excellence. We will also review how risk information can be used to inform decision making in the field, and whether its success or otherwise can be measured in a meaningful way.

The book is filled with DRA case studies, tips, practical examples, solutions and lessons from the field which share ideas for what to do (and what not to do).

Case studies

You will find eighteen detailed case studies in these pages – several in each chapter. These address the application of SRA/PRA/DRA, the findings and learnings from a broad array of industries including aviation, sports, health services and construction, as well as 'blue light' settings. Where applicable, information from these cases is published with the consent of those quoted, the investigating body or the families of the injured or deceased. The book and its companion website (see www.routledge.com/cw/dynamic-risk-assessment) also contain information already in the public domain which has been published by the HSE and licensed under the Open Government Licence v1.0 (www.nationalarchives.gov.uk/doc/open-government-licence/).

Tips

The authors have considerable experience of using DRA in work settings. Throughout the text there are 'Tip' boxes in which you will find a generous serving of their tips, which you can use in your own workplace. These are tried and tested, and come highly recommended to you.

Many of the tips are accompanied by examples of DRA 'at work'. These were collated during our research for this book and come from sources including Formula 1 motor sport, fire and police services, construction, care work and public health services. Each example comes with a suggested solution for providing safe and healthy conditions in a dynamic working environment.

These case studies, tips, examples and solutions are spread relatively evenly throughout the book's chapters. Together, they comprise the golden nuggets of information which crystallise or summarise major points in the text.

We look forward to building upon these ideas and sharing new experiences in future editions of this book. We will also try to support those interested in DRA through our book's companion website www.routledge.com/cw/dynamic-risk-assessment.

There, you'll find a host of useful additional materials, including:

- articles and papers of interest
- a list of useful websites

- court reports of interesting legal cases
- incident reports
- regulators' guidance.

And you can also keep up to date with risk management news and solutions by following us on Twitter @CRS_tweets.

We look forward, in the following 144 pages, to sharing with you and developing your appreciation of the powerful tool that is dynamic risk assessment.

About the authors

Stephen Asbury is the Managing Director of Corporate Risk Systems Limited, a leading international HSE training and consulting organisation. He has authored around fifty books, articles and technical papers on safety and risk management. Stephen has worked in a variety of senior risk management roles in employment and in consulting in over sixty countries on six continents in a career which spans almost thirty years in the following sectors: construction, polymers, heavy engineering, oil and gas, pharmaceuticals, insurance and across a broad range of technical consultancy assignments at high-value assets.

Stephen is a Chartered Fellow of the Institution of Occupational Safety and Health, a Professional Member of the American Society of Safety Engineers and is registered by the Society for the Environment as a Chartered Environmentalist. After college, his first qualification was in law.

In his leisure time, he enjoys theatre, scuba diving and F1 motor sport.

Corporate Risk Systems Ltd
Clay House
5 Horninglow Street
Burton upon Trent
Staffordshire
DE14 1NG
United Kingdom
www.crsrisk.com

CORPORATE RISK SYSTEMS
SAFETY HEALTH AND ENVIRONMENT

Edmund Jacobs is Head of Fire, Health and Safety for the Ministry of Justice (Head of Profession) and has nearly twenty-five years' experience in the field of occupational health, safety and well-being, across a wide range of large and complex organisations. He has practical experience of the application of DRA in many organisations, providing advice and training, and he is a regular speaker at conferences in this subject matter.

Edmund has a master's degree in Occupational Safety and Health and is a Chartered Fellow of the Institution of Occupational Safety and Health (IOSH), as well as a Fellow of the Institute of Directors (IoD). He is an Honorary Member (and former Chairman from 2005–2008) of the London Health and Safety Group and a Director and Vice Chairman for Quo Vadis Trust, which is a charity that is dedicated to supporting people recovering from mental health issues. When time allows, he enjoys walking and going to the theatre, and he is a life-long Chelsea supporter.

Elite OHS Solutions Limited
Apt 2260
Chynoweth House
Trevissome Park
Truro
TR4 8UN
www.eliteohssolutions.com

Acknowledgements

Stephen Asbury

In almost thirty years in the health and safety profession, I have seen too many lives changed by injuries – for workers and their families. Some years ago, a very good friend of mine suggested that I share my experiences of working to prevent this by writing more widely and, since 2006, I have been doing just that. This is my fourth book and, like the earlier ones and all of the papers I've penned, it seeks to engage you, the reader, in doing something to eliminate or control work risks. Workers commonly have big brains, and if we are to succeed in preventing life-changing events, we must engage those brains and encourage their use. There is so much more to health and safety than filling in the paperwork.

My thanks go to Edmund Jacobs for asking me to write with him. He told me that I was a good writer and that he had enjoyed my earlier books. He is far too kind, and I hope my contribution has justified his confidence. Paul Richardson has once again been my illustrator, and I feel he has brought a touch of humour where it was appropriate, as well as sensitivity where sad or tragic events are depicted.

My wife, Susan, has supported me again and I apologise for the dinners that went cold while I was 'just finishing this paragraph'. And for my poor contribution to the housework (again). She is my rock, and I'd find my journey so much more difficult without her. Thank you, darling.

My special thanks to all of the contributors and interviewees I spoke with during the groundwork for this book, in particular Deborah Clements at Vodafone, Andy Stevenson at Sahara Force India F1 and Dr Craig Stenhouse at Burton NHS Trust. Captain 'Sully' Sullenberger did a great job with US1549, and my parts of this book are dedicated to him.

My thanks to Steve and Sheree Martin for helping me to arrange interviews, and to the team at CRS for covering my role while I worked on this project, in particular Ian Cliffen, Ros Stacey, Richard Ball, Stewart Clarke and Jonathan Backhouse. You're always a pleasure to work with, and very supportive colleagues.

Edmund Jacobs

Having worked in some very challenging environments, I recognise the importance of protecting workers and the public, while still getting the job done. Whilst serious injury or loss of life can never be eradicated, we need to continue to strive to ensure that we make the best possible decisions to prevent unnecessary harm. It is often too easy to forget the lives that have been profoundly affected by health and safety failings. During my research for this book, I found it humbling to speak with Margaret Aspinall, who is one of the relatives of the ninety-six people who died in the Hillsborough disaster. Her personal story was so moving and I would like to thank her for her trust, support and contribution.

To Stephen Asbury, my thanks – we always said that we would collaborate, and I am pleased that we have done so on a subject I am passionate about. I would like to express my gratitude to the many people who saw me through this book by providing support and useful advice; particularly Mario Weick, Paul Hampton, Jim Parrott, John Arnold, Simon Pilling, Bill Fox, John Norton-Doyle and Malcolm Brown. Thank you to all those I have not mentioned, but am just as grateful to.

Above all, a special thank you to my wife, Janette, for your encouragement and support – without you this would not have been possible; and of course my two sons, John and James, for supplying me with copious amounts of tea and chocolate biscuits, to keep me going.

Chapter 1

Introduction to dynamic risk assessment

In any moment of decision the best thing you can do is the right thing, the next best thing is the wrong thing, and the worst thing you can do is nothing.

(Attributed to Theodore Roosevelt 1858–1919,
26th President of the USA)

Introduction

Thinking is vital in our lives; we need to do this to consider options, plan future activities and to make decisions. We like to think of ourselves as rational human beings, who think logically in order to make the best decisions that fit in with our desired goals.

Dynamic risk assessment (DRA) is a scaffold which aids in the decision-making process under conditions which are fast paced and changing. So just what is DRA? We're going to explain this, and describe its concepts; psychological influence from naturalistic decision-making research; and its evolving applications in many different types and size of workplaces, across all employment sectors.

Over the last fifteen years, the base of knowledge on DRA has expanded considerably and, as a result, we see an evolving understanding of DRA and application of its use. However, we also refer to, in the authors' opinion, misunderstandings, myths and misconceptions concerning its utility and purpose, as against its effectiveness. As well as explaining what the relationship of DRA is to traditional 'predictive' risk assessment, we will explore these misunderstandings, myths, applications and misconceptions to clarify and to inform the reader.

This chapter will also explain the assessment of risk in dynamic environments and review its roots, explaining how it emerged from the recognition primed decision making paradigm in naturalistic settings, which is based in the field of psychology. Chapter 2 will discuss the emergence of DRA and how it has cut across non-emergency service occupations. Chapter 3 explains how predictive risk assessment is undertaken in slow time and applied in fast time in the dynamic environment. The concept of the three levels of risk management – strategic, predictive and dynamic model – will be discussed in more detail in Chapter 4, along with the organisational structures in which DRA should exist in order to be effective. Chapter 5 will consider how DRA can be embedded within your organisation management systems and Chapter 6 will explore the benefits of DRA and how it can be used to improve organisational service delivery. Chapter 7 will explain the importance of underpinning DRA with training and applied learning. Finally, Chapter 8 will share some insightful examples of how to learn from organisations that utilise DRA and of the value of feedback and debriefing.

We will raise and answer questions; and challenge your preconceived beliefs, and when we are done, we will leave you with some tips that will help you to integrate DRA within your organisation in a sustainable and efficient way. We hope that this will compel you to constructively review and challenge your organisational approaches to risk management at all levels.

The topics that will be covered in this chapter include:

- introduction
- a potted history of DRA
- the psychological base of decision making
- the origins of DRA
- the origins of naturalistic decision making
- the characteristics of DRA
- concepts of DRA
- myths and misconceptions
- tips, examples and solutions.

For organisations considering DRA as a risk-based approach to dynamic decision making, or those organisations that already use this methodology and are looking for improvement, this book will provide you with a wider understanding and necessary techniques to take you forward. Our concepts will be underpinned by a further seven insightful chapters to enable the practitioner, the manager, worker, people in the decision-making chain and customer-facing staff to determine the validity of DRA as a useful, recognised and systematic process for inclusion within their overall approaches to risk. It will also assist senior executives in their understanding and recognition of the tangible value of adopting a risk-based and streamlined approach to aid decision making at the strategic level as well as at the predictive and dynamic levels.

A potted history of DRA

DRA was founded within the fire service in the mid 1990s (led by the London Fire Brigade). Following several high-profile fire-fighter deaths in service during the early 1990s, the Health and Safety Executive (HSE) served a number of statutory improvement notices on the Fire Service. Two notices were served on the London Fire Brigade following the deaths of two fire-fighters at Gillender Street in 1991. The HSE then served a third notice after a fire-fighter's death at Villiers Road in 1994. In addition, two more improvement notices were served on Hereford and Worcester Fire Brigades in 1993, following the deaths of two fire-fighters. Together, these highlighted the need for better risk assessment in relation to the systems of command and tactical fire fighting (Flin, 1996). In recognition of these HSE requirements and recommendations, the Fire Service refocused its attention on managing the risks to its fire-fighters who routinely work in hazardous situations. This refocusing led senior fire service managers and subject matter experts in the fire discipline to develop and agree DRA as a concept for use in the field. The Home Office subsequently published in 1998 the Fire Service guide *Dynamic Management of Risk at Operational Incidents* and *A Guide for Senior Officers Health and Safety and Fire Service Operations Incident Command*. This was issued to all operational personnel in England and Wales.

The original Home Office guidance had defined DRA (albeit now superseded) as:

> The continuous assessment of risk in the rapidly changing circumstances of an operational incident, in order to implement the control measures necessary to ensure an acceptable level of safety.
>
> (HM Fire Service Inspectorate, 1998)

During the 1990s the then Health and Safety Manager at London Fire Brigade played a pivotal role in the proposal, adoption and promotion of the DRA concept, and was an important supporter in the subsequent implementation of DRA in the UK Fire Service.

Prior to the adoption of the concept of DRA in the Fire and Rescue Service, the service had carried out assessments for many years operating in highly hazardous and constantly evolving situations. It is argued that the need to make rapid assessments to inform tactical decision making was nothing new; this is born out of naturalistic decision research (Fire Service Manual, 2008). This assertion equates to the concept of situation awareness and assessment generally referred to earlier as 'size-up'. This process of situation awareness is undertaken by the Incident Commander, who is then responsible for issuing orders during an operational incident (Home Office, 1981, cited in Tissington & Flin, 2005).

Dugan (2007, cited in Lusk, 2008) highlights the distinction between size-up and situation awareness. He says that size-up relates to the method fire-fighters use to evaluate the fire conditions in a building, the resources available or the number of victims affected by the fire. He says situation awareness involves not only gathering these facts, but also understanding what the information means. The Department for Communities and Government (DCLG, 2013) refers to situation awareness as:

being vigilant for personal safety and the safety of team members, being observant and able to identify and react safely to new or unexpected hazards, particularly when working without supervision.

Research by Tissington and Flin (2005) found that the introduction of DRA led to a significant cultural change within the Fire and Rescue Service, with risk now being central to the way crews are managed by their fire commanders.

There was a significant shift from the Fire and Rescue Service situation size-up to now including a risk-based approach to decision-making methodology, following the introduction of DRA. Moreover, situation awareness has become a key component of the rapid identification response for all commanders operating in real-time situations (Fire Service Manual, 2008).

Outside of the Fire and Rescue Service, organisations adopting DRA interpret and use DRA in a variety of different ways. For the authors, DRA is a continuous cognitive (mental) assessment of risk in a changing (dynamic) environment which aids decision making to provide an acceptable level of safety. This will undoubtedly consist of a myriad of influencing human factors and complex issues, which the book will cover in great detail in later chapters.

Psychology-based decision making

It is also important to consider the psychological basis of decision making in relation to risk. The subject matter of decision making has been of particular interest to psychologists. Research has been conducted for many years on how individuals and groups make decisions. Over the years, many psychologists have suggested that humans have dual processes for reasoning and problem solving (Evans & Over, 1996; Sloman, 1996). Stanovich and West (2000) named these 'system 1' and 'system 2'. System 1 is said to use implicit knowledge and would be utilised in fast decisions based on intuition, and processes information quickly. It is argued by many to be an unconscious/ preconscious reasoning that is often influenced by previous experience, memories and emotions. System 2 uses explicit knowledge and would be used in slow, conscious, controlled decisions and for logical, rule-based decisions. System 2 requires more mental effort, whereas system 1 is automatic. Stanovich and West argue that system 1 is more prone to biases, as we utilise our own beliefs, personal experiences and intuitions to solve problems. Whilst system 2 is more effortful, it is still not immune to the influences of beliefs, biases and emotions.

Whilst investigating judgement and decision making, psychologists have also been interested in faulty decision making, as these errors can provide us with insights which may aid us in the use of better decision-making strategies. We like to think that we make rational decisions to reach set goals, and in an ideal world we would have lots of time to consider all the information and available options. When individuals lack specific knowledge about risks or are under time constraints, they have to make inferences based on salient information in order to judge how likely the risk is to occur, and its likely severity. This type of inferential judgement is better known as heuristics and concerns strategies that are used to reduce cognitive load – mental short-cuts that allow us to make quick judgements based on information from a similar situation that we can readily call to mind (system 1). People often refer to these as a 'rules of thumb'. These heuristic decisions help us to make decisions without using too much brain power. Whilst these strategies can be useful, they can lead to severe and systematic errors or biases in judging and deciding on risks (Slovic, Kunreuther & White, 1974). Some of these types of errors (biases) are listed in Table 1.1.

We are all subject to these types of biases in our everyday lives. These biases can lead to serious mistakes in judgement, and being aware of these types of error can help us to make better decisions, particularly if this is reinforced in training as a part of learning and a continuous improvement cycle.

Of course, not all decisions will be made by one individual; often the person may be in contact with senior managers, back-office support staff or co-workers. This may help with decision making by pooling experience and knowledge and allowing others to assist (e.g. to manage back-up resources such as police, equipment, other staff). One would like to think that by using the joined-up approach, several experts with a wealth of experience would consistently achieve a better performance (make better decisions). However, research has suggested that this may not always be the case, with Stasser and Titus (2003) finding that when people gather to discuss issues/problems they often make inferior decisions. This is due to individuals not disclosing all of the information available to them. They also found that where there is a lot of information present, individuals find it hard to remember it all. Stasser and Stewart (1992) argue that one of the main issues is that the group tend to spend more time discussing known, pooled information and do not consider asking individual group members to contribute more information or question the information present. There are other biases in groups that can lead to poorer decisions (Table 1.2).

Asch (1940; 1948) argued that individuals tend to conform to group beliefs/views/decisions (labelling this phenomenon social conformity). He carried out a study in which individual participants were placed in a room, with people who they believed to be fellow participants, but who were in fact actors. When presented with questions the group had to decide on the correct answer; the group deliberately chose an incorrect answer and tried to convince the individual participant that the obviously incorrect was indeed correct. They found that people conformed to group decisions for several reasons, such as: they knew the answer was wrong but did not want to dissent from the group; they originally believed the group view to be incorrect but were eventually convinced that it must be right because so

Table 1.1 Error biases

Type of bias	Description
Availability bias	When people judge a risk to be more or less likely based on how readily they can recall to mind previous examples of the risk (e.g. from personal experience, a film or book).
Representation bias	When people base their judgement of a new risk on how much it resembles a known risk that they have encountered.
Anchoring	When individuals rely on an initial piece of information (starting point) when making further decisions. Individuals adjust their decisions around the starting point (the anchor).
Overconfidence	Where an individual subjectively measures their judgement to be higher than it objectively is. Overconfidence can also have the effect of individuals not thinking through all the options, leading to poor decisions and outcomes that shock the individual (Cohen, Freeman & Thompson, 1998).
Hindsight bias	Where individuals have a tendency to review past actions/events by exaggerating their ability to predict the outcome (prior to hindsight knowledge). They suggest that they knew what the outcome would be all along, when in fact they did not.
Confirmation bias	Tendency for people to search for new information by unwittingly selecting information that is consistent with their existing attitudes, beliefs or views.
Framing effect	Where individuals can reach different conclusions using the same information, depending on by whom or how the information is presented.
Risk compensation	This is when individuals tend to take greater risks when they perceive that the level of safety has increased.
Near-miss bias	Dillon & Tinsley (2005) found that where events turned out successfully through luck rather than through successful decisions, people often failed to recognise the dangers and rejected that the result had occurred through luck. People attribute the success to their skill and tend to become accepting of risky decision making, making riskier decisions in future.
Illusory optimism	This is where individuals remain unrealistically optimistic for the future, even when they have negative information. People tend to update their beliefs using positive information and often fail to assess negative information. The view that 'it will work out in the end' (Sharot, Korn & Dolan, 2011).
Myside bias	Where individuals may fail to search for or ignore evidence that is against a belief/view they already have.

Table 1.2 Group biases

Type of bias	Description
Group polarisation	This is where groups tend to make more extreme decisions than an individual. If the individual's initial decision is to be cautious and other group members hold this view, then they may be collectively more cautious than they would originally have been (prior to group influence). People also have a tendency to believe in what the majority think, because with so many people agreeing – they must be right.
Groupthink	This is when individuals struggle to achieve consensus within a group. They may withhold information or not voice their true beliefs or opinion in order to facilitate group harmony. Janis (1982) argued that there were three major causes of poor thinking: 1) overestimation of the group 2) closed-mindedness and 3) pressure towards uniformity.

many others believed this; or they seemed to truly believe the group view all the way through the study.

Emotion has also been shown to affect our judgement of risk (Janis & Mann, 1977). Bargh and Chartrand (1999) argue that we have an automatic and effortless appraisal response to almost everything we see (stimulus) and that it registers as either good or bad. Emotions can cause people to act in an illogical manner (on instinct) and disregard their own safety (Dodson, 2004, cited in Lusk, 2008), as in the case of fire-fighters who may ignore their operational responsibilities and personal risks to go to the rescue of a colleague in danger.

In some circumstances the general public opinion can also contribute to our perception of risk taking (Trimpop, 1994). Risk takers can be seen as heroes (a police officer who jumps into a raging river to save a child's life), fearless (stunt artists and racing car drivers), and society often honours and rewards this kind of behaviour. However, if people avoid risk (no matter what the outcome) they can be labelled as cowards, weak or dishonourable (e.g. army deserters) (Soane & Chmiel, 2005).

There are also other important factors that affect risk perception, and one of these is personality. Zuckerman and Kuhlman (2000) found evidence that with six different risk activities (smoking, drinking, drugs, sex, driving and gambling) risk taking was related to three different personality traits or factors: impulsive sensation seeking, aggression and sociability. Other researchers have also had similar results. For example, West and Hall (1997) found that sensation seeking, aggression and social deviance are particularly related to road traffic accidents. Even in unknown situations or unfamiliar activities individuals who are 'high sensation seekers' will evaluate risks as being lower than do 'low sensation seekers', who are more likely to be anxious when considering undertaking a risky activity (Horvath & Zuckerman, 1993).

The origins of DRA

Michael Crichton, author of *Jurassic Park*, said:

> If you don't know history, then you don't know anything. You are a leaf that doesn't know it is part of a tree.
>
> (Crichton, 2013)

Our adaptation of the above quote is, if you don't know history, you'll never know the world!

So, with history in mind, let us trace back to the 1980s, where much of the research behind DRA emerged. The origins of DRA were based on decision-making research which also looked at how experienced personnel such as Incident Commanders made decisions in their natural field environment. Naturalistic decision making (NDM) research was heralded as a new paradigm and is the study of how domain experts make decisions in the field, such as in fire-fighting and aviation operations.

The origins of NDM

DRA evolved from the recognition primed decision making (RPDM) model of NDM. So where did the NDM paradigm come from? NDM research comes from the study of decision making in psychology, and has over time superseded the discipline of Classical Decision Making (CDM), which dates back to the 1940s and 1950s and can even be traced back to the eighteenth century. CDM had distinguishing features that focused on predictions based on having a range of decision options from a selection of alternative choices. The decision maker would need to weigh up all the available information when selecting the best action. For the best option, researchers Lipshitz, Klein, Orasanu and Salas (2001) found that the NDM doctrine requires field experts to size-up the situation to determine the nature of the problem and, on the basis of previous recognition, to determine an appropriate course of action.

In essence, NDM researchers and practitioners have typically found that, in the moment of decision making, the field expert would not seek to be exhaustive in their search for the optimum solution. Instead, they would seek to choose an option that would provide an adequate solution to the problem. Thus, NDM research is concerned with understanding how people make decisions in the field pertinent to their own environment. NDM research has tended to focus on command-and-control type organisations in naturalistic settings. According to research by Klein and Klinger (1991), commanders asserted that their decision-making aim in field settings was to find an effective and workable solution to the situation, rather than seeking to deliberate on decision choice for optimisation. Incident commanders involved in the research would not delay, so as to avoid exacerbating the situation.

Commanders tended to use mental strategies that rely on prior experience for dealing with similar situations, such as with mental matches, and, where appropriate, to use modifications and adaptations.

Klein developed the RPDM model. Within the NDM framework RPDM defines how field experts use their prior knowledge to inform decision making. This illustrated in Table 1.3.

It is well documented that recognition-primed decision-making strategies based on prior experience to inform decision making response were used by emergency services (Lipshitz et al., 2001). These included tank platoon leaders, commercial aviation pilots and offshore oil installation managers. The research found that the selection of first decision choice reduced in line with experience. The challenge for personnel in these types of occupations is the importance of gaining continually relevant experience to improve and maintain decision-making skills.

The characteristics of DRA

We have described some of the concepts of DRA and the rapid cognitive processes involved in assessing risk, and its strong link to decision making. Additionally, this mental process involves considering the benefits of proceeding with a task and weighing them carefully against the risks involved, which should be underpinned by organisational standard operating procedures. It also seeks to equip personnel with a consistent and common approach to assessing risks in the field. DRA is inextricably linked to predictive risk assessment carried out as part of careful planning prior to the activity taking place. The characteristics of DRA founded upon predictive risk assessment have distinct features by virtue of being applied in situations that involve:

- unpredictability or unforeseen risks
- situations where the risk environment changes rapidly
- shifting and competing goals
- dynamic and continually changing conditions requiring continual assessment
- high-risk stakes
- pressure to act swiftly with either incomplete or inaccurate information
- personnel required to make risk judgements in the field.

Concepts of DRA

Dynamic risk assessment is a term used when the *environment or situation* in which risk arises is dynamic, rather than the risk itself. DRA is central to decision making and is concerned with thinking before you act, rather than acting before you think. Jacobs (2010a) highlights that assessing risk on the

Table 1.3 Characteristics of naturalistic decision making

1	Ill-defined and ill-structured goals and tasks
2	Uncertainty, ambiguous or missing data
3	Multiple players and teams
4	Experienced decision makers
5	Time stress
6	High stakes and risks
7	Uncertain, dynamic environments
8	Shifting and competing goals
9	Action feedback loop (real-time reactions to changed conditions)
10	Organisational goals and norms

Source: Klein and Klinger (1991)

spot involves one or more conditions of pressure, constraint of time and dynamic environment/situation, and requires the worker to have a good understanding of risk assessment (based on training) and its importance in decision making. For example, in a fire-fighting context, whilst it is necessary for an effective risk assessment to be undertaken at an operational incident, it would not be practical for a written assessment to be undertaken at the scene. Instead, the incident commander would carry out a mental assessment and the decision action and feedback loop would be time-stamped and communicated via radio (FSEB, 2003). Organisations should consider decision making in these terms when setting up or overhauling their risk management systems.

To embed the methodology of DRA within an organisation, it will require DRA to be established and integrated within the risk management and governance structure supported within an occupational health and safety framework (this will be discussed in detail in Chapter 6). Notwithstanding the need for continual improvement, the Fire and Rescue Service has led the field on DRA, aligning the RPDM methodology within the dynamic management of risk framework to support decision making and achieve operational delivery. This mental evaluation model, illustrated in Figure 1.1, consists of an on-going assessment and risk-based decision making prior to, during and at closure of the activity.

Myths and misconceptions

We highlighted earlier that the understanding of DRA is developing and we believe that further research in the field outside of the command-and-control risk environment is necessary to improve knowledge and understanding. Whilst advancement is being made, it is not a silver bullet. It is crucial that we don't lose sight of the objective and how it should be applied. With that in mind, we'll tackle head on some of the myths and misconceptions that may hinder or undermine your understanding of DRA concepts.

The following are some common misconceptions that we have come across.

Initial attendance stage of incident

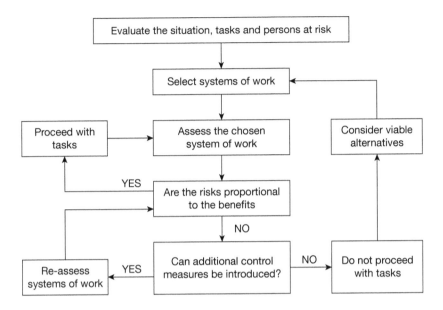

Developmental stage of incident

As the incident develops, re-evaluate the situation, tasks and persons at risk. Apply above model to take account of any new hazards and introduce control measures as necessary to allow existing or new tasks to proceed. Halt tasks completely if the risk outweights the benefits to be gained.

Closing stage of incident

Maintain the process of task and hazard identification, assessment of risk, planning, organisation, control, monitor and review of the preventative and protective measures.

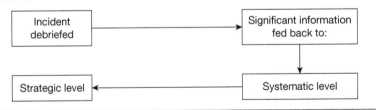

Figure 1.1 Three stages at an operational incident

Source: Home Office (1998)

'DRA is just an alternative to activity-based risk assessment'

It is important that DRA is not seen by the employer (the duty holder) as a substitute for the legal duty to undertake a predictive risk assessment for their staff's activities, or relied upon in isolation. Duty holders need to consider the risks carefully and identify and implement proportionate controls prior to staff carrying out their duties, and if there are more than five employees they need to record their significant findings.

Remember, if your risk management processes aren't right, your DRA won't be right either.

'My staff do DRA, so I'm covered!'

Just saying 'my staff do DRA', without proper reference to predictive risk assessment, which is a line management responsibility, is not in itself a legal defence. DRA is not something done in isolation; it is an integral part of your overall risk assessment process. An organisation cannot rely on DRAs without formal recognition of a risk management system. Line managers need to set the risk-based framework in which their staff operate and assess risk in dynamic situations and how it fits into the overall health and safety system.

No risk assessment, no compliance!

If correctly integrated, DRA methodology can play an effective role in an organisation's risk management system.

'DRA – it's a tick-box, back-covering exercise'

DRA must not be seen as a bureaucratic exercise. Like the predictive risk assessment undertaken at your workplace, if it is done correctly and applied appropriately in the field, it can aid staff to make sensible decisions.

DRA also provides the means for a consistent and formalised system, which can be evidenced in the field, that controls are working (or otherwise). It can also become a valuable means of learning and feeding back within the overall risk management system.

'DRA is dealing with dynamic risk'

It is important to focus your attention on the dynamic situation in which the risk arises and the key relationship it has with your predictive risk assessment, and the safe systems of working, rather than dynamic risk per se.

'I've done a DRA and I'm not going to do my job because it's too dangerous!'

It is important that when carrying out tasks, we do not become overly risk averse and use DRA as an excuse for not doing our job. Health and safety has been targeted in certain quarters of the media to either trivialise or sensationalise stories. It is important that health and safety is not used as an excuse for not doing your job. Staff need to focus on carrying out their duties in an efficient and effective way in as safe a way as the conditions allow. Undertaking an assessment of risk in the field should provide you with a means for choosing the right course of action, which is supported by your organisation. Thus, risk assessment facilitates the completion of the job, instead of restricting it.

Case study 1.1 Road safety: assessing the risk

If you were a child in the UK anytime between 1970 and 1990, you will probably remember being taught the Green Cross Code. A televised campaign supported the initiative. The Green Cross Code standardised an approach to road safety, and encouraged children to practise their skills (under supervision). This involved looking right, left and right again; and keeping looking and listening while they crossed. This enabled a child to gain their own experience in how to cross roads safely (see Figure 1.2).

At a basic level, the principles of DRA can be exemplified in the way you learn to cross the road as a pedestrian. Thinking about the risks and how you

Figure 1.2 The Green Cross Code

can safely get to the other side, what controls are available? We often don't think of risk in these terms, due to our experience in crossing roads – it becomes intuitive over time. Many of the everyday risks in crossing a road are predictable and we can therefore use previous knowledge to help us put safety controls into place. For example, using a zebra crossing, or crossing away from parked cars, and so on. Every time we cross a road we gain knowledge and this information forms part of a mental framework or structure which helps us to interpret, process and organise information. Whilst the risk of being knocked down when we cross the road is always present, the environment changes and circumstances will vary every time we cross, even if we cross at the same location at the same time every day. Nothing is or stays exactly the same. For example, the weather conditions may change; there will be drivers who are familiar with the area, which brings its own challenges, and those who are not regular users of this road; the speed of cars will change; likewise the condition of the road; drivers could be distracted by mobile telephone calls or they could be rushing to an appointment. Not to mention your own state of mind, your risk appetite, judgement on a given day and familiarity with the road and so on.

In summary, when crossing a road you make a quick cognitive assessment of the road hazards, both before you cross and whilst crossing, until you safely get to the other side. You apply controls relevant to the circumstances and, over time, when sufficient experience has been gained (largely through our childhood) in crossing roads, the process becomes intuitive and greater reliance is placed on past experiences. These principles can be readily applied to a work setting through strategic, predictive and dynamic levels of risk assessment, which will be discussed later on.

Case study 1.2 Welly-wanging DRA and its incorporation into case law

This case study concerns *Cornish Glennroy Blair-Ford* v *CRS Adventures Limited* (England and Wales High Court (Queen's Bench Division) Decisions, 2012). On 19 April 2007, the claimant, Mr Glenn Blair-Ford, suffered a spinal injury in an accident which occurred in the course of throwing a Wellington boot backwards through his legs at an outdoor pursuits centre as a part of an organised mini-Olympics. The claimant sustained permanent tetraplegia, and the case presented was upon the issue of liability.

At the time, Mr Blair-Ford was reported to be a 6-foot tall, sporty, 15-stone, 40-year-old man. He was head of design and technology at a college and had a history of playing rugby and throwing the discus. He was also a keen cyclist, riding 10 miles to and from school two or three times a week. Mr Blair-Ford would also do a 5-mile run once per week, would swim regularly (on average three times a week) and used light weights at home.

Mr Blair-Ford and other teachers and pupils took part in the welly-wanging competition at the school's outdoor event at the defendant's outdoor pursuits centre. Welly-wanging involves throwing a Wellington boot backwards, in this case between your legs. When Mr Blair-Ford came to throw the welly, his head would have been low down by his legs, and his hands went through his legs with great force, throwing the welly high into the air, but it did not gain much distance. It was during this motion that he toppled over and fell onto his head and neck, rather than toppling forwards onto his face and chest.

As a result of the accident, the claimant alleged that the defendant, CRS Adventures Limited, was negligent in its adoption of an unsafe throwing method and that, as a result, there was a foreseeable risk of injury. The claimant believed that, had the risks been explained to him, he would have prepared to put his hands out in protection against the impending fall.

Figure 1.3 Welly wanging

Some of the issues considered in *Cornish Glennroy Blair-Ford* v. *CRS Adventures Limited* were:

1   Was a formal health and safety management system in place?
2   Were formal risk assessments in place for mini-Olympics as a whole?
3   Was there an assessment of risk before the specific activity?
4   Company's previous safety record.
5   Was the activity dynamic in nature and the risk foreseeable?
6   Were staff suitably trained in risk assessment and job role?
7   Previously known incidents/risks associated with the event.

Summary of Judge's conclusion

The Court was satisfied that the defendant had acted reasonably, as shown in the Judge's conclusions.

1   Defendant had a formal risk assessment for mini-Olympics as a whole.
2   Good safety record (no other incidents involving welly-wanging).
3   Specific risk (formal) assessment was not required due to the unforeseeable risk.
4   Judge satisfied that a DRA was acceptable (due to variable factors involved).
5   Variables involved, e.g. uneven ground, unusual throwing.
6   Based on a DRA, no steps were needed to modify the method or to provide any specific warnings.
7   The claim was dismissed, with judgement entered for the defendant.

A copy of the judgement is available on this book's companion website (www.routledge.com/cw/dynamic-risk-assessment).

This is an important case, as the courts have now formally recognised that DRA plays a key role in an organisation's occupational health and safety management system. However, DRA is not a stand-alone process and must be considered as part of a wider risk management system.

Case study 1.3 Work at height in the Police Service

As in the Fire and Rescue Service, police officers sometimes operate in extremely hazardous environments and often operate alone as 'single' units. They may be the first to arrive at the scene of an incident, and one of the many risks is falls from height.

Historically, the Police Service was exempt from the requirements of the Health and Safety at Work etc. Act 1974 (HSWA), as its members were regarded as office holders appointed by warrant and seen as sworn officers of the Crown rather than employees. Police authorities and chief officers were asked to 'voluntarily' co-operate with the spirit of the legislation, whereas police support staff were covered under the HSWA, from the outset.

The government removed this exemption at the request of the Police Service (Home Office, 1998), with the subsequent introduction of the Police (Health and Safety) Act 1997 (PHSA), which applied Part 1 of the HSWA to police officers. This important piece of legislation was quickly followed by the Police (Health and Safety) Regulations 1999, which required forces to comply with other relevant health and safety statutory provisions. These regulations have brought police officers, special constables and cadets under the control of health and safety legislation.

During these significant changes in safety legislation for the Police Service, several incidents occurred within the Metropolitan Police Service.

Two of these incidents included serious accidents during the chasing of suspects at height. In the course of his duties, Police Constable (PC) Sidhu fell through a roof in October 1999, as did PC Berwick in May 2000. PC Berwick sustained serious injuries, whilst PC Sidhu lost his life in the line of duty on 24 October 1999. Following these incidents the Metropolitan Police produced an aide memoire in memory of PC Sidhu covering a list of top ten high-risk activities, which was distributed to police officers. The Metropolitan Police Authority (2004) aide memoire provided safety guidance on key safety issues that officers should consider (Figure 1.4). These risks included:

- chasing suspects (including working at height)
- bombs and suspect objects
- fires
- traffic collisions
- searching people
- hazardous materials

- firearms and lethal weapons
- domestic incidents
- mental disorder
- water rescue.

Figure 1.4 Police pursuit in the high street

TIP

Before your organisation decides to adopt DRA, consider its applicability to the nature and risk profile of your business, as well as its operational environment.

Example: An outdoor instructor organised a swimming event that involved each participant having to dive into the water. One of the participants couldn't swim and had to be rescued from the water.

Solution: When organising events that require a level of skill, make sure the participants are able to fulfil any skill requirement.

TIP

Adopting DRA methodology requires an understanding of the relationship to predictive risk assessment and of measures to embed the process.

Example: A male suspect wielding a knife whilst being chased by police officers (wearing stab-proof vests) in attendance. A police officer broke from his work to join the chase (without a stab-proof vest) and pursued the knifeman on a narrow canal bank and was subsequently fatally wounded.

Solution: Always wear personal protective clothing issued by your employer to protect against the hazard involved in the activity.

TIP

You need to link core values, service delivery and expected business outcomes with the need to communicate risk expectations to your people in the context of risk management.

Example: A member of staff working for the emergency services was the first to arrive at the scene of an incident and was anxious that he did not have the necessary equipment to commence a rescue, and had a member of the public begging for him to take action.

Solution: Employers need to make it absolutely clear what is expected of their staff, through policy, guidance, procedures and risk assessment.

TIP

You need to adopt a consistent and integrated approach to training that is delivered by knowledgeable people who have proven experience of effectively implementing DRA principles and practice within organisations.

Example: An employee started a new job as a case worker and, prior to his training, was required by his boss to visit a client in their home. Given his lack of training and information, the worker did not know what to do when confronted by a violent client.

Solution: As part of induction and risk assessment for the proposed employee activities, training and information must be provided to employees who are faced with significant risk in order to provide them with appropriate skills and preparedness to mitigate the risk.

TIP

Training and greater accountability need to be given to line management in recognition of their key roles in the process.

Example: A police officer radioed the control room to ask for back-up whilst following a person acting suspiciously in an alleyway. The control room staff sought advice from a senior manager on the number of back-up personnel available, as resources were stretched. The senior manager made the decision that back-up personnel were not required and over-ruled the control room officer. When this was relayed back to the person in the field they made the decision not to put themselves in harm's way, which led to the individual being informally reprimanded for their decision.

Solution: Training in the application of DRA should be targeted and given to line managers, and not just to front-line staff in the field, as line managers have the responsibility for predictive risk assessment.

Chapter summary

In each of our chapters, we'll give you an opportunity to check your knowledge before you move on.

1 What does RPDM stand for?
2 What is the conceptual distinction between predictive risk assessment and dynamic risk assessment?
3 In which environments is DRA typically applied?
4 What is the definition of DRA?
5 Is it a legal defence to say, 'I have done my DRA and I am legally covered'?
6 In which industry sector did DRA originate?
7 What does NDM stand for and why is it important to the DRA paradigm?

8 What is hindsight bias and what issues does this bias pose when making decisions?

9 Name three characteristics associated with DRA.

10 At what stage would you undertake a predictive risk assessment, and why?

Chapter 2

The emergence of dynamic risk assessment

Dynamic risk assessment is a process of common sense decision making to enable officers and staff to manage the inherent day to day risks in policing; from normal response policing to fast time major incidents. At high-risk fast time incidents, individual officers and operational staff will carry out subjective assessments of hazards and take appropriate and immediate actions to manage the hazards and control the risk. Dynamic assessments are made based on guidance and training. This process does not replace the requirement for the formal risk assessment of tasks and activities by the employer.

Metropolitan Police Service (2010)

The Myth Busters Challenge Panel provides a mechanism for anyone (whether on behalf of a company or organisation, or as an individual), who receives advice or is told that a decision has been taken in the name of health and safety that they believe to be disproportionate or inaccurate, to challenge that advice.

Health and Safety Executive (2013a)

Introduction

It's very important that you 'get' risk assessment. Wherever you are in the world, and regardless of the type or function of your organisation, it is a moral, a financial and/or a legal requirement these days. And we don't mean goofing around with the paperwork and completing the forms, nor risk assessments which result in 'disproportionate' decisions (Health and Safety Executive, 2013b). What we do mean is making a careful consideration of the significant risks to your employees and others affected, taking specific and particular actions to prevent their injuries or ill-health, and documenting the significant findings so that they can be replicated and subjected to an improvement process. Everyone we have spoken to during our research for this book has told us that risk assessment concerns the quality of decision making during a number of critical times. This book will explain how to do this, at three key and critical levels:

1   Strategic, high level – may be commitments or policies (usually made in slow time).
2   Predictive, systematic, planning level – may be generic and/or site-specific but informed by the above (again, usually made in slow time).
3   Dynamic and on the spot – informed by the above, but decided very nimbly (need to be made in fast time).

You'll find our model for properly assessing risks, including in a dynamic way, in Chapter 4. We explain how to integrate this proven approach into your own organisation's structure and your safety and health culture. Don't go thinking this is a static model or a carved-into-kryptonite approach. Oh no! We explain extremely clearly how an intentionally continuous loop of assessing, implementing, reviewing and improving can be achieved. Some call this the *Deming Wheel* or the *PDCA Wheel*. Whilst the characteristics of their fast-time operations and the need to make risk-based decisions on the spot at the scene have led emergency and higher-hazard services such as police forces, armed forces, commercial pilots and paramedics to adopt this methodology, you'll see, as we have in our research, that all types of organisations can take similar approaches. The case studies in this book prove beyond reasonable doubt that managers and health and safety professionals should take account of our dynamic risk assessment (DRA) model. But, for a moment, we are ahead of ourselves.

The origins of risk assessment

This chapter covers the origins of risk assessment, and thus DRA. We explain how it emerged in legislation and in regulators' guidance. You'll have read in Chapter 1 about how and why DRA has moved across the emergency services and higher-hazard environments to arguably lower-hazard, non-emergency environments in other organisations and sectors. This chapter reviews how risk information can be used to inform decision making in the field and whether its success or otherwise can be measured and thus appreciated. A good starting point is to review the legal requirements in your territory for assessing risks.

TIP

Review the applicable legal (and any other) requirements for risk assessment. As well as the law, these requirements may include your own health and safety policy commitments, the expectations of your sector and your insurers' conditions of cover. Check the policy details of this latter if you are unsure. Some of the established occupational health and safety management frameworks require precisely this review, such as clause 4.3.2 of BS OHSAS 18001:2007 (British Standards Institution, 2007).

Example: Pearson plc is the world's leading education organisation as well as owning the *Financial Times*. It has over 50,000 employees based in almost half of the countries of the world. Pearson has operations in 96 of the world's current total of 196 countries. Very few organisations have a presence in so many legal jurisdictions. Pearson's Group Health and Safety Policy requires all group businesses to identify and document the legal and other standards which apply to them. Another group standard requires an (at least) annual review by the senior management of each business that the standards identified are being met. Group audit provides assurance to the board of directors of the proper application of these standards.

Solution: We recommend that you conduct a comprehensive review of the legal and other applicable requirements in your own organisation, and write these requirements down. The list of useful websites on our companion website (www.routledge.com/cw/dynamic-risk-assessment) provides links to all health and safety legal requirements, including Acts, Statutory Instruments (Regulations) and legal guidance in the UK. This can be reviewed, and bespoked to reflect your own organisation's needs, which will of course depend upon its sector and work activities. Time prohibits us from researching and presenting the requirements for the world's current other 195 countries, but you'll be able to research your own country/ies in a much more focused manner.

Origins of health and safety, and risk

Hammurabi code

Some say the earliest reference to health and safety legislation emerged in around 1772 BCE in one of the oldest deciphered writings, known as the Hammurabi code (Asbury & Ball, 2009). Hammurabi was a Babylonian king who enacted a code of 282 laws written on 6-foot stone and clay tablets. The code provides punishments set on the well-known scale of 'an eye for an eye, and a tooth for a tooth'. For example, it set out the punishments for a builder who has constructed a house which collapses, killing the son of its owner; the son of the builder must also be put to death.

History of risk

Asbury (2013) provides a summary of the fascinating history of risk in his book about management systems auditing. He says that you would not have to go back in time many years for much of the modern clarity of approach and measurement to be lost. A well-educated individual a thousand years ago would not recognise the number '0', and would probably not pass a basic mathematics test. Five hundred years later, few would do very much better. Without some form of measurement, or some numbers, risk was a matter of 'gut feeling' or superstition.

Asbury says that the 'power of numbers' arrived in West in the early thirteenth century, when a book entitled *Liber Abaci* appeared in Italy – a wholly hand-written fifteen volumes by Leonardo Pisano (commonly known as Fibonacci).

Fibonacci is best known for a series of numbers which provided the answer to the problem of how many rabbits will be born during the course of one year from one pair, assuming that every month each pair produces another pair, and that rabbits start breeding at the age of two months. The answer is 233, and the twelve month-end totals for the year would be 1, 2, 3, 5, 8, 13, 21, 34, 55, 89, 144, 233. Each successive number is the sum of the two preceding numbers, and if one number is divided by the next, the answer is approximately 1.6. This ratio features in nature (e.g. in shell spirals, leaves and flowers) and in architecture (e.g. the General Assembly Building of the United Nations in New York). Playing cards are similarly proportioned. The Fibonacci series also features in the book and later film *The Da Vinci Code*, where the dying Jacques Saunière leaves a code for Robert Langdon to decipher. Fibonacci identified the 'power of numbers' in the West for the first time, but using them to assess risk remained many years distant.

Bernstein (1996) commented on the development of humans' understanding of risk over the last millennia:

> What is it that distinguishes the thousand years of history from what we think of as modern times? The answer goes way beyond the progress of science, technology, capitalism and democracy . . . The revolutionary idea that defines the boundary between modern times and the past is the mastery of risk: the notion that the future is more than a whim of the gods and that men and women are not passive before nature. Until human beings discovered a way across that boundary, the future was a mirror of the past or the murky domain of oracles and soothsayers . . .

Bernstein (1996) gives an interesting account of this history, suggesting that:

> The ability to define what may happen in the future and to choose amongst alternatives lies as the heart of contemporary societies.

Hazard and risk

A modern definition of hazard is 'the potential for harm'. The word 'hazard' is said to derive from the Arabic word for dice – *al zahr*.

There are many definitions of risk, and some are given in Table 2.1. The word 'risk' is said to derive from the early Italian *risicare*, which means 'to dare' (Asbury 2013). To dare implies the freedom to choose and, as a result, possibly to fail.

Dice is a game of luck – of pure chance – of potential for harm, of hazard. Whilst lots of things have potential for harm (*al zahr*), managers can choose – to dare or not – how and when to respond to hazards. This choice influences the probability of the harm occurring and the consequences, should the harm be realised.

This 'daring' to participate actively in the business environment includes choices for managers.

Risk assessment

Although the UK has had health and safety legislation since 1802, the term 'risk assessment' has been in common use only since the inception of the single European Market in 1993. The European Framework Directive EEC 89/391 was implemented in the UK as the Management of Health and Safety at Work Regulations 1992 (MHSWR) (SI 1992 No. 2051) and amended in 1999 (SI 1999 No. 3242). Regulation 3 of MHSWR made 'risk assessment' a commonly known and commonly used term in workplaces and other settings.

It is, however, often said that the Health and Safety at Work etc. Act 1974 in the UK provided some impetus to assess risks. This Act, also referred to

Table 2.1 Some definitions of risk assessment

HSE (1997)	. . . the combination of the severity of harm with the likelihood of its occurrence . . .
Boyle (2002)	[A] combination of the hazard and the loss and, in any given set of circumstances, risk takes into account the relevant aspects of both.
Fuller and Vassie (2004)	. . . the chance of a particular situation or event, which will have an impact upon an individual's, organization's or society's objectives, occurring within a stated period of time.
BSI (2007)	. . . combination of the likelihood and consequence(s) of a specified hazardous event occurring.
Wikipedia (2012)	. . . the potential that a chosen action or activity (including the choice of inaction) will lead to a loss (an undesirable outcome). The notion implies that a choice having an influence on the outcome exists (or existed). Potential losses themselves may also be called 'risks'. Almost any human endeavour carries some risk, but some are much more risky than others.

as HSWA, HSW Act or HASAWA, is the primary piece of legislation covering occupational health and safety in Great Britain. Section 2(2)(a) of HSWA requires employers to provide and maintain plant and systems of work which are, so far as is reasonably practicable, safe and without risks to health. With hindsight, it was probably not possible to develop effective and lawful safe systems of work without some attempt to identify:

- the risks to health and safety in the organisation, and
- which of those risks called for the greatest attention by management.

In short, risk assessment is fundamental to answering the question of what was required of employers to achieve health and safety compliance.

In the timeline of the emergence of risk assessment, the original edition of the COSHH Regulations (Control of Substances Hazardous to Health Regulations 1988, SI 1998, No. 1657) was the first health and safety legislation in the UK to change the historic pattern of providing a precise definition of what was required to comply with the law. COSHH was the first to tell employers to assess their own business risks and to take whatever steps were necessary so far as reasonably practical to eliminate or mitigate the risks associated with hazardous substances.

Other 'risk assessment' laws followed, most notably MHSWR, Regulation 3 of which requires assessments to be 'suitable and sufficient'. A few examples of some other health and safety laws requiring a risk assessment-based approach include:

- Manual Handling Operations Regulations
- Health and Safety (Display Screen Equipment) Regulations
- Regulatory Reform (Fire Safety) Order
- Control of Noise at Work Regulations
- Control of Asbestos Regulations
- Health and Safety (First Aid) Regulations.

Five steps to risk assessment

In 1999, HSE published its guidance document to assist employers to undertake risk assessment in a series of five steps. Revised twice since, the document called 'Five steps to risk assessment' (HSE, 2011), says in its Preface:

> A risk assessment is an important step in protecting your workers and your business, as well as complying with the law. It helps you focus on the risks that really matter in your workplace – the ones with the potential to cause real harm. In many instances, straightforward measures can readily control risks, for example ensuring spillages are cleaned up promptly so people do not slip, or cupboard drawers are kept closed

to ensure people do not trip. For most, that means simple, cheap and effective measures to ensure your most valuable asset – your workforce – is protected. The law does not expect you to eliminate all risk, but you are required to protect people as far as 'reasonably practicable'. This guide tells you how to achieve that with a minimum of fuss.

The five steps to risk assessment are summarised below:

Step 1   Identify the hazards
Step 2   Decide who might be harmed and how
Step 3   Evaluate the risks and decide on precautions
Step 4   Record your findings and implement them
Step 5   Review your assessment and update if necessary.

Readers can download a copy of 'Five steps to risk assessment' indg163 (rev 3) from this book's companion website at www.routledge.com/cw/dynamic-risk-assessment.

In 2012, HSE placed thirty-three examples of template risk assessments for shops, motor vehicles, offices and other locations on its website www.hse.gov.uk/risk/casestudies/. You may find it useful to review these if you are a manager or are otherwise engaged to advise on health and safety in these work environments.

Management system standards and BS OHSAS 18001

Asbury (2013) provides a history of management systems standards (MSS) for business control dating to ancient China in the fifth century BCE. Many MSS were developed in the 1980s which followed the well-known Plan-Do-Check-Act cycle developed by Deming, Shewhart and others.

BS 5750 was an early standard for quality assurance proposed to the International Standards Organisation in 1979. It was based on a series of US Defense standards, including MIL-Q-9858 in 1959. This was adopted by NATO in 1969, revised to BS 5179 in 1974, and revised to BS 5750 in 1979. ISO 9000 was published in 1987; today, through several revisions, it is called ISO 9001:2008. It is presently being revised again in light of experience of use and to reflect the harmonisation of all MSS owned by ISO which is established by *Annex SL* (ISO, 2012); it is expected to be re-issued in 2015. Likewise, BS 7750 on environmental management systems was adopted internationally as ISO 14001. This is also presently being revised in line with experience and *Annex SL*.

BS OHSAS 18001 is a British Standard for occupational health and safety management systems. The words used in the standard are *OH&S Policy, Planning, Implementation and Operation, Checking,* and *Review* (from BS,

2007 – see Figure 2.1), but these are all aspects of the recognition that senior management must *plan* what needs to be achieved in both quantitative and qualitative terms, ensure an effective implementation, *do* in accordance with the developed plan, conduct verification and measurement, *check* what has been done, and *act* to consolidate standards, or alternatively to initiate an improvement where this has been revealed to be necessary ('better next time').

BS (2007) says (clause 4.3.1):

> The organization shall establish, implement and maintain a procedure(s) for the on-going hazard identification, risk assessment, and determination of the necessary controls. The procedure(s) . . . shall take into account routine and non-routine activities . . .

ILO-OSH 2001

Figure 2.2 illustrates the occupational safety and health (OSH) framework developed in 2001 by the International Labour Organisation (ILO). ILO aims to create worldwide awareness of the dimensions and consequences of work-related accidents and diseases; to place OSH on national and international agendas; and to provide support to national efforts for the improvement of national OSH systems and programmes in line with relevant international labour standards.

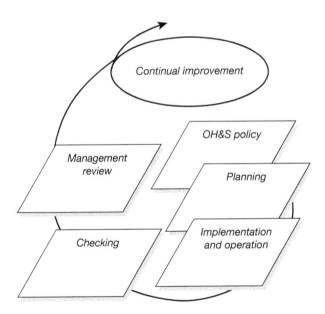

Figure 2.1 Policy, plan, implement/operate, check and review

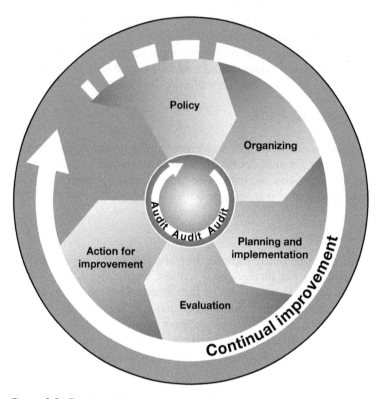

Figure 2.2 Continual improvement cycle

ILO-OSH 2001 clause 3.10.1.1 says:

> Hazards and risks to workers' safety and health should be identified and assessed on an on-going basis.

Risk assessment and the emergency services

Firefighters acknowledge that their work will occasionally put them in hazardous situations and they are willing to accept some risk to their personal safety in order to protect the communities that they serve.

(Henry McLeish, George Howarth and Alf Dubbs, ministers with responsibility for fire service matters at the Scottish Office, Home Office and the Department for the Environment (Northern Ireland) in HM Fire Service Inspectorate (1998))

As we introduced to you in Chapter 1, the Fire and Rescue Service was the first organisation to consider its own approach to assessing risks, especially at incidents. Fire Service Circular 5/1995 (Scottish Fire Circular 4/1995) discussed this, and a Joint Committee on Fire Brigade Operations (1996) has issued *Guidance on the Application of Risk Assessment in the Fire Service*.

Chapter 1 and Figure 1.1 presented the three stages of an attendance at a fire or emergency incident during which time different matters will have to be considered as the incident progresses. The three stages are:

1   Initial – upon arrival at the scene
2   Development – as the incident develops
3   Closing – during the closing stages of an incident.

Chapter 5 will explain the 'Safe Person' concept as described therein.

This book will consider the application of DRA in Fire Service case studies later; but for now we wish to start to highlight to you how DRA either might have been used, or has been used, in some other sectors and other organisations. Each case study also includes learning, tips and ideas that you can use in your own organisation.

Case study 2.1 Fatality at Glenridding Beck, 2002

Three high school staff accompanied a party of twelve Year 8 pupils to Glenridding, in Cumbria, UK for a weekend of adventure and activities. The group also contained three primary school-age children, including the deceased, Max Palmer. The deceased's mother was one of the adults accompanying the visit.

On the Sunday morning, the party went to a pool in Glenridding Beck to do an activity called 'plunge pooling'. This involved jumping 4 metres into a rock pool in a mountain stream and swimming to an exit point. Parties from the school had done the plunge pooling activity a number of times on past activity trips.

On the weekend of the incident, the weather was cold and wet. The stream was fast running and the water very cold. Immediately after jumping in, Max was seen to be panicking and was unable to get out. The leader jumped in to rescue him, but after a few minutes was overcome by the cold and left the pool. Max's mother, who was helping out on the weekend, then jumped in, but was also overcome by the cold and had to be rescued – she too almost drowned. The pupil who (probably) saved her from the water was also affected by the cold. Both he and the mother were airlifted to hospital and treated for hypothermia.

Max was washed over the weir at the exit of the pool. He was pulled from the beck approximately 150 metres below the pool, but was pronounced dead at the scene. The investigation of this tragedy by the UK's health and safety regulator, HSE, and the police showed:

- serious errors of judgement by the party leader in planning and leading the activity – this was judged to be the main cause of the tragic incident
- shortcomings in checking procedures
- shortcomings in the school's arrangements for educational visits
- misunderstandings as to health and safety responsibilities.

There had been other similar, previous occurrences. In October 2000 two schoolgirls were drowned while river walking in Stainforth Beck, UK.

TIP

It is naturally good practice to review your own health and safety procedures in the light of reported incidents and developments in local, national and international best practices.

GLENRIDDING BECK

Figure 2.3 Plunge pooling

The incident in context

Under health and safety law in most territories, including the UK:

- employers are responsible for ensuring that there is effective management of health and safety
- people such as school managers and teachers are responsible for the health and safety of pupils both when they are authorised to be on the school premises and when they are on authorised school activities elsewhere.

Besides the overriding need to put the health and safety of children at the very top of the education agenda, effective health and safety management systems are needed to:

- meet legal requirements
- assure parents that the school is good at managing risks
- give staff the confidence that they can rely on well-planned health and safety systems when carrying out their responsibilities.

Elements of successful health and safety management

Successful health and safety management has five key elements (HSE, 1997):

1   establishing policies
2   organising the health and safety management system
3   planning, assessing risks, setting standards and implementing procedures
4   monitoring performance
5   review.

Investigation and report into the events at Glenridding Beck

Experience shows that accidents usually result from several factors coming together to cause harm.

The incident report is an interesting read, with much learning for health and safety practitioners, as well as for those organising trips and events. We have summarised the main contents of the report; in its thirty-eight pages, it:

- summarises the very detailed investigation by the regulator and the police into the tragic incident
- shows that the chain of events leading to the tragedy began long before the fateful weekend
- shows how compliance with existing guidelines and good practice prevents such chains of events from developing, and identifies some new issues

- makes clear that such prevention relies on having effective health and safety management systems and guarding against individual and institutional complacency
- demonstrates the importance of pupil involvement in organising safe and successful educational visits.

The full report into the tragedy at Glenridding Beck is available from this book's companion website (www.routledge.com/cw/dynamic-risk-assessment).

Authors' comment: strategic–predictive–dynamic

We observe the following characteristics in relation to our 3-Level Risk Management Model™ in Chapter 4. Whilst the activities had been undertaken before by children from this school, under supervision by some of these school staff, conditions in mountain rivers can and do change considerably over short periods. Heavy rain can quickly change a shallow, warm stream into a raging, freezing torrent. No amount of policy and strategic assessment identifying the importance of child protection, nor slow-time risk assessment conducted in the weeks prior to the weekend, can replace proper review of the factors on the day.

Any person proposing to lead an activity such as plunge pooling would need to understand the significance of a number of variables (hazards) which could give rise to danger (risks). These variables include:

- water depth
- water flow
- water temperature
- water currents
- air temperature
- rocks and other physical features
- the ability and make-up of the particular group.

DRA in non-emergency environments

Events at Glenridding Beck in 2002 provided an early, but by no means solitary, example of the value of DRA in non-emergency organisations.

Examples of where our research has identified DRA in use (either in name, or often in spirit) include the flowing sectors:

- aviation
- construction
- estate agency
- Formula 1 motorsport
- health service provision

- highway maintenance
- local authorities
- offshore oil and gas exploration and production
- road safety
- security and door supervision
- space exploration
- sports and recreations, including mini-Olympic 'welly-wanging'.

These and other cases you'll read about in this book and elsewhere provided the springboard for DRA to be adopted elsewhere.

TIP

Familiarise yourself with the approaches to assessing risks in your own organisation. These are the three main types:

- generic risk assessment – the risks inherent in the activity (written)
- activity or site-specific risk assessment – particular risks associated with the site or activity, e.g. water depth, ease of exit, difficulty of rescue (written)
- DRA – the risks at the time, taking account of, for instance, the changing conditions and the fitness of the party to undertake the activity (physical, mental and attitudinal).

The 'myths' highlighted in Chapter 1 show that none of these is a substitute for another – the types are interconnected.

TIP

Risk assessments provide any line manager's fundamental intelligence on health and safety. It is not enough to 'have done this loads of times before', or to be 'very experienced' in this type of work activity. Risk assessments are the route to identifying the right control measures in the present – not the last time, nor the next time – and not an end in themselves. Risk assessments need to be fit for their purpose (suitable and sufficient). If having the risk assessment paperwork is seen to be more important than properly implementing the control measures, the system is very probably poorly focused, and probably too complex or poorly understood. Go back to the basics and start again.

TIP

Senior managers in all organisations need to know what is going on. They can (and should) appoint competent health and safety advisors to assist in the discharge of their legal and moral duties, but the accountability still remains with top management.

Example: Likewise, governors and parents in education settings should be prepared to ask searching questions about the educational objectives and management arrangements for forthcoming educational visits and the outcomes of completed visits. Even if the visit has a limited educational purpose, the leader still has responsibility to organise the activity to the best of their professional ability. There can be no lowering of standards because it is a 'fun trip'. While visits (in this instance) may have certain primary activities which may be seen as the 'highlight', it is always important to have a viable 'Plan B' to provide alternative activities of educational value in case the primary activity is undeliverable on the day. 'Plan B' is a viable alternative activity, not an emergency plan. A similar standard of risk assessment should be applied to 'Plan B' as to the main activities.

TIP

It should be clear who, in any organisation, has specific health and safety responsibilities. The extent and limits of their functions should be clearly set out.

Solution: Informed parental consent is essential. This requires good communication with parents as follows:

- It is good practice to hold a meeting for parents before outdoor and residential visits.
- It is good practice for a governor to be invited to attend parents' meetings for visits.

TIP

The risk assessment should identify the staffing required to run a visit safely. Staffing should never be decided just by a simple numerical calculation of ratios, as has sometimes been the case. Activity leaders must be careful not to create unrealistic expectations and should not allow participant pressure to cloud their judgement.

Example: The 'sudden panicker' phenomenon is increasingly being recognised as one of the risk factors in outdoor drownings. In the context of the major case study in this chapter, this should be considered in the risk assessment and emergency plan.

Solution: People should not jump into water outdoors without having first swum in it to assess the conditions for themselves.

TIP

DRA should continue throughout the day/activity to take account of all and any changing circumstances.

TIP

Make sure that the employees of your organisation are competent in the area of risk assessment at all three levels, as applicable to their role.

Solution: Check 'claimed' qualifications. Note too that the fact that someone has participated in or managed an activity before does not, in itself, mean that they are competent. Of course they can read this book, though our experience is that those who need these competences the most seek them the least. So, as a manager or a health and safety practitioner in your organisation, you may need to push such training at those responsible for the safety of others. If you need a specific solution, look at www.routledge.com/cw/dynamic-risk-assessment – the authors deliver the training personally.

TIP

It is best practice for generic and activity and/or site-specific risk assessments to identify specific criteria which can be used as reference or calibration points for the subsequent DRA.

Solution: These might (for example) be absolute in their terms, e.g. 'we don't do the activity if the water is above this level'. Identify additional precautions, e.g. 'wet suits are essential if the water is colder than N°C'; 'buoyancy aids are mandatory if the depth of the water exceeds NN centimetres or if the group includes poor swimmers'; and/or 'helmets are required if there is a risk of contact with rock'. Set a minimum age or skill level required to undertake the activity safely in particular conditions, e.g. 'participants under 18 require additional supervision'.

Case study 2.2 Captain Sullenberger lands on the Hudson River

On 15 January 2009, there was a remarkable news story which reported the emergency landing of an Airbus A320 (an 'A320–214', for those interested in the details) operating US Airways flight 1549 into the Hudson River, New York, saving the lives of all 155 passengers and crew aboard (Sullenberger, 2009).

> The flight lasted just a few minutes, but so many of the details are rich and vivid to me. I have a clear recollection of how my body felt – this heightened sense of alertness – as I taxied to the end of the runway, went through my checklist and got ready to go. And I recall the moment the plane lifted into the air and, just three minutes later, how I would need to return to the runway. I've just described my first solo flight. It was June 2 1967, and I was sixteen years old. As I worked to safely land Flight 1549 in the Hudson, almost subconsciously, I drew on those experiences.
>
> (Sullenberger, 2009: 1–2)

Captain of the aeroplane that day was Chesley 'Sully' Sullenberger. He had developed a passion for flying from watching jets above his childhood home in Denison, Texas, from the age of five. After school, he enrolled in the US Airforce Academy in 1969. He was selected as one of a dozen other freshmen to participate in a glider programme, and by the end of the year he was an instructor pilot. In his graduation year, 1973, he received the 'Outstanding Cadet in Airmanship' award and was officially recognised as the class 'top flyer'.

Following graduation, he was commissioned as an officer. He served as a US Air Force fighter pilot from 1975 to 1980, flying the McDonnell Douglas Phantom. In that time, he advanced to become a flight leader and a training officer, and finished his service with the rank of captain.

After his service, Captain Sullenberger was employed by US Airways from 1980 until 2010. At the time of his retirement, he had more than 40 years and 20,000 hours of flying experience, as well as experience of participation in a number of aviation accident investigations, including Pacific Southwest Airlines flight 1771, and US Air flight 1493. He was active with his union, serving as chairman of a safety committee within the Air Line Pilots Association.

> Flight 1549 wasn't just a five minute journey. My entire life led me safely to that river.
>
> (Sullenberger, 2009)

On 15 January 2009, Sully responded to a bird strike which knocked out both engines on his plane in less than 60 seconds, to determine the fate of his aircraft 'N106US' and her passengers. This is shown in the summarised timeline in Table 2.2 (Sullenberger, 2009: 323, 328–40).

Table 2.2 Timeline of US Airways flight 1549 (call-sign 'Cactus 1549')

15.24.54	Cactus 1549 is cleared for take-off from LaGuardia Airport, Queens, New York.
15.27.10.4	The word 'Birds' is recorded on the in-cockpit recorder.
15.27.11.4	2750 feet, 219 knots. Bird strike [sound of thump/thuds].
15.27.28	In-cockpit 'Get the QRH [quick reference handbook]. Loss of thrust on both engines.'
15.27.32.9	3020 feet, 185 knots. 'Mayday, mayday, mayday' on radio transmission. 'We're turning back towards LaGuardia.'
15.27.42	Air traffic control (ATC) gives return heading of two two zero.
15.28.10	Captain Sullenberger tells ATC 'We may end up in the Hudson'.
15.28.19	1560 feet, 204 knots. Finishing turn in south-southwesterly heading.
15.28.46	1260 feet, 195 knots. Passes over George Washington Bridge.
15.28.49.9	Captain Sullenberger questions the availability of Teterboro airport (5 miles west in New Jersey).
15.29.11	1050 feet, 190 knots. Cabin public address system to crew and passengers 'This is the captain. Brace for impact.'
15.29.53	ATC offers Newark airport as a possible landing site 'off your two o'clock in about seven miles'.
15.30.43	125.2 knots. Emergency landing in Hudson River. Everyone survives.

Note: Times are Eastern Standard Time (EST), based on the clock used to timestamp the recorded radar data from Newark ASR-9.

Figure 2.4 Captain Sullenberger safely lands US Airways flight 1549

Authors' comment: strategic–predictive–dynamic

We observe the following characteristics in relation to our 3-Level Risk Management Model™ shown in Chapter 4.

Strategic level

Airlines fly modern aircraft and engage experienced pilots and first officers, both of whom are competent to fly the plane.

Air incident investigations are taken very seriously, including those of near misses, so that learning can be shared within the industry.

Predictive level

Every flight has a flight plan, and progress against this is maintained through communications with air traffic control.

Flight plans anticipate flying conditions, and aircraft and crew are prepared accordingly.

Dynamic level

Weighing up and implementing the options and alternatives, including Teteboro and the Hudson River, in fast time, based on 40 years' and 20,000 flying hours' experience – including experience from the pilot's first flight – saved the day.

TIP

The role for rules: Make sure that employees know the rules for whatever it is that they do. Selective compliance with the rules has led to many incidents.

Example: Some drivers are a bit over-confident. They're not sure that all the rules of the road apply to them.

Solution: Try to make sure that employees know that rules are there to prevent anarchy (in this example, on the roads).

TIP

Expect the unexpected: Make sure your staff know about 'Black Swans' – low-probability, high-consequence events that may affect them.

Example: Airline pilots take off and land again and again without incident.

Solution: Make sure that your staff know what to do in emergency situations. Train and drill them, so that they know what to do. Your strategic and operational risk assessments will inform them.

TIP

Look out for confirmation bias: People find things that confirm their point of view.

Example: Stephen is passionate about timekeeping and being on time. His wife says that matters for clients, but not when they go on holiday, as it's not necessary. She says 'life is not a checklist' – BUT the Asbury family always gets to the airport on time.

Solution: Recognise that people wish to have their views heard, but be ready with a well-researched answer (particularly if the point of view has been long held, and you have time to do so).

TIP

Situational awareness: Make sure your employees are always situationally aware. This means that they can always create and update an accurate, real-time model of what's going on.

Example: A fighter pilot tried an aggressive manoeuvre too close to the ground and didn't have enough space to complete it. Looking at a barometric altimeter set to give readings above sea level does not give you the height above ground, which was 3,000 feet above sea level on that day. As pilots say, he lost the picture until it was too late to correct it.

Solution: Always know what you do and don't know. You need to know how your judgement may be affected by circumstances.

TIP

Reporting incidents: Make sure that all employees know that you want all incidents reported so that you can investigate them. Even no-injury incidents matter, as they provide opportunities for learning.

Example: At a company Stephen used to work for, management pressed for all incidents to be reported. Over a five-year period, substantially fewer lost-time injuries were sustained, as the organisation had really learned about hazards and how to control them.

Solution: Don't give rewards for low incidents. If you do, people tend to under-report. Instead, reward better, more accurate reporting. This tends to encourage the reporting of incidents.

Case study 2.3 Lessons from Formula 1

Andy Stevenson has worked in Formula 1 since 1991, always with the same Silverstone-based team, although it has changed its name a few times over the years. Today, it is called Sahara Force India Formula One Team. Previously, it was Jordan Grand Prix, Spyker F1 and Midland F1. Andy is the Sporting Director of the race team, and his role takes him to every race around the world each year.

Stevenson (2013) describes several scenarios where the strategic plans of the team and individual race strategies have been amended in fast time by events.

The Team has been owned by several individuals and organizations over the years, each with its own reasons and objectives for involvement in the sport. We have a long-service workforce based in our original factory at Silverstone which adapts to those objectives, and the budgets available to us. For a long time, we have punched above our weight. We hire the best drivers we can and build the best car we can within the requirements of the regulations of Formula 1.

Each Formula 1 season has nineteen or twenty races around the world. This year [2013], our prediction of the performance of the tyres in the prevailing conditions at each circuit is critical to race success. The tyre supplier for this season is Pirelli, which provide each of the eleven teams with details of the two tyre compounds, one harder and one softer, we're to use 7–10 days in advance of each race. The more time we have, the more data we can crunch on the compounds and how they will react. We use any information available to us such as the performance at last year's race.

Our next race is at the Hungaroring (Hungary); the long-range weather forecast is for a very hot race – maybe the hottest F1 race ever. We also have a new type of tyre to manage. We can predict the performance of these tyres through our modelling and data-crunching, but we'll not see the performance for real until we run our two cars on the Friday, two days before Sunday's race.

We run in the two 90-minute practice sessions on the Friday evaluating each of the two tyre types, as well as testing any other upgrades to the cars. We gather data throughout, and on Friday evening we make a team decision on our race strategy. This is informed by our IT and with our drivers' input for 'feel'. In the final practice session on the Saturday, we'll rehearse our qualifying and race strategies. We have to decide our chosen race strategy before qualifying commences.

Planning on our pit-stop strategy for each race concerns deciding how many times and at what intervals to change our tyres on each car will be critical to the outcome. A pit stop takes 20–25 seconds, depending upon the length of the particular pit lane. We must keep to the pit lane speed limit each time our cars come to the pits. Our target time to change the four tyres on the car is 3 seconds, and while we have done this in less time, doing it safely remains a major priority to me.

In Malaysia earlier this year [Malaysia Grand Prix, Sepang International Circuit, 24 March 2013], we were running well in the race. We had developed a revised type of captive wheel nut which we felt would be quicker to remove and replace the wheel as the nut does not need to be relocated by the wheel gun, and cannot be dropped. When the system faulted and the nut would locate onto the thread in the race, the decision to withdraw the car was taken 'on the corner' – by the mechanic changing the wheel. He knows best what can be done, and not. He followed our procedures to the letter, and I was extremely happy, despite the disappointment, about this. Safety has to come first.

We try to cover every eventuality in advance, and we are extremely procedure driven. For each race, we'll develop several scenarios – the exact number must remain confidential. If there is an incident in the race such as the deployment of a safety car by the F1 Race Director to control the race during specified conditions, we'll already have many of our pre-planned scenarios and optional strategies available to us. Key to this is the position of each car on the track. If we wish to make an earlier pit stop, we will have only seconds to react, to tell the driver, and have the pit crew and tyres ready. We keep our actual latest reaction times secret, but if say we have 45 seconds before our car passes the entry to the pit lane, that's how long we have to decide what to do. We have to decide in these seconds whether to stack the cars in the pit lane [have one car wait behind the other], or have the second car make an extra lap prior to stopping. Drivers don't like being stacked, even for a few seconds, and even when we tell them it's the fastest strategy for them. And if it goes wrong for the first car, it affects both cars.

In Germany in 2007 [European Grand Prix, Nurburgring, 22 July 2007], we could see the high risk of rain in the race. This was not one of our strongest seasons [the team finished 2007 in 10th place]. We took a gamble of starting our car from the pit lane on wet tyres. We were thus in 22nd (and last) position. The time it took for the rain to arrive and the sheer amount was heavily under-estimated by everybody but us. Within one lap, our decision to start on wet tyres placed our driver Markus Winkelhock driving in his first race in the lead at his home Grand Prix after everyone else pitted for wet tyres or stayed out on dry ones. This was the first time that Spyker had led a Grand Prix. Only the temporary red flagging [stopping] of the race due to the weather conditions and to recover cars in unsafe conditions that had aqua-planed off the track prevented us from finishing better than we did.

In each race, we have a team of eight on the pit wall. I decide what can be done in the time and within the rules. The race strategist and the engineer running each car make the decisions about our race. Only the Team Principal or the Chief Operating Officer can over-rule those decisions.

At Silverstone 2013 [British Grand Prix, Silverstone, 30 June 2013], our senior team decided to do whatever we could to get a podium [top 3] finish as it was our home race. That decision guided our decision-making for the whole weekend. We were very focused on it. We had the right car, the right tyres, and the right drivers. Only a late safety car from which we could not recover stopped us from being successful, as it gave other teams the time to stop and change their tyres. But we were pleased to bring both cars home in the points at our home race.

Figure 2.5 Sahara Force India F1 team

> After each race, we have a full debrief with everyone making decisions and working within those decisions in the room. They are very organised, and very well documented. We usually start 20 minutes after the race while information is fresh, but long enough to allow everyone to calm down. The Chief Engineer chairs the meeting. Our drivers take turns at alternate races to speak first, and then anyone can contribute. We all have information to contribute. If someone says 'we should have done this or that on lap 13', we have the tools to test such ideas and provide an accurate answer, such as 'that would actually have been 4 seconds slower'. We don't focus on that though, as we generally don't learn from it; and we don't allow Monday morning quarterbacks [next day suggestions that someone should have done this or that]. We see our improvement as the main objective.

Authors' comment: strategic–predictive–dynamic

We observe the following lessons from Formula 1 which may be useful to other organisations. It was interesting that Stevenson should identify so strongly and positively with our DRA model shown in Chapter 4.

1 Working many (race) options in advance provides a series of ready-made choices for implementation on race day (predictive level).
2 Empowered devolution of decision making to the front line means that decisions are made at the place where the information is available (strategic level to devolve, dynamic level where decision made).
3 The right decision 'then' is – by definition – the right decision 'now' (strategic level).
4 Good predictive risk assessment can also deliver an excellent outcome (predictive level; it is clear that a DRA is not always required).
5 Decisions must be timely. A decision needed in 45 seconds is of no value delivered in 46 seconds (dynamic level).
6 Debrief everyone in the team after the event to ensure team and individual learning. There is no tolerance of 'Monday morning quarterbacks' (to inform future assessments and decisions).

Chapter summary

As you'll have seen in Chapter 1, we give you an opportunity to check your knowledge before you move on to the next chapter.

1 Where in the law (in the UK) might you find the earliest legal requirement for a predictive risk assessment?

2 Why should a health and safety practitioner identify the legal and other requirements relating to health and safety in their own territory/country?

3 What is OHSAS 18001:2007? For an extra mark, which clause of the standard requires identification of legal and other requirements?

4 What is ILO-OSH 2001?

5 What happened in Yorkshire, UK in 2002 which provided an insight for others outside the emergency services that an approach to risk assessment beyond a pre-planned assessment might be beneficial?

6 Which of the emergency services is attributed with introducing the DRA concept?

7 What are the three main stages of a fire service operational risk assessment?

8 What are the six 'lessons from Formula 1'?

9 How can any manager confirm that the employees of their organisation are competent in the area of risk assessment at the levels applicable to their role (i.e. strategic, predictive and/or dynamic)?

10 What should any manager do if they believe that completing the predictive risk assessment paperwork has become more important than properly implementing the control measures?

Chapter 3

The application of predictive risk assessment in the field

You can't make decisions based on fear and the possibility of what might happen.

(Michelle Obama, 2013)

Introduction

In Chapter 2 we looked at the three levels of risk management, which were strategic, predictive and dynamic, and the key role these play in providing a framework to support an organisation to achieve its objectives. In this chapter we discuss the principles of risk assessment undertaken in slow time, and applied in the field in fast time, utilising the principles of the five steps to risk assessment. We explore ways in which organisations can enable their workers to be safe in dynamic situations and to make effective, risk-based decisions in the context of achieving operational and wider objectives.

Here we summarise the areas of consideration. The topics that will be covered include:

- predictive risk assessment (PRA) undertaken in slow time
- employer responsibilities
- relationship of PRA to dynamic risk assessment (DRA)
- the role of human factors
- tips, examples and solutions.

In order to understand the principles of risk assessment and the purpose it serves a business, it is helpful to get back to basics to clarify your understanding of hazard and risk. In everyday parlance hazard and risk are generally interchangeable – Chapter 2 provided the history of these terms. For example, people often say 'policing is a risky occupation', when it fact what they really mean is that policing can be a hazardous occupation. Hazard and risk need to be considered in the context of organisational meaning, their interface with workplace activities against the backdrop of your health and safety management system.

PRA undertaken in slow time

Throughout this book, we talk about predictive (traditional) risk assessment, which is required to be undertaken by law. The purpose of this assessment is to ascertain whether a risk arising out of a workplace hazard is acceptable or requires further controls to reduce it to an acceptable level. Organisational PRAs need to be presented in a way that is systematic, logical and coherent. The adequacy of an assessment, should it ever be called into question, would need to satisfy the regulator (e.g. HSE) that it was suitable and sufficient, and this would ultimately be determined by the courts. Weick, Hopthrow, Abrams and Taylor-Gooby (2012) usefully highlight that our awareness of risk as individuals will be different, in terms of how we reason, our thinking process and filtering of information that will inform our judgement and, ultimately, our decisions. Whilst we cannot eliminate individual biases or differences we can work within a risk framework which utilises a systematic and consistent process of risk assessment on how the activity should be safely undertaken. Organisations should therefore arrange their systems in such a way that line managers can confidently and competently fulfil the task of risk assessment in slow time, enabling workers to understand how it can be applied in the field. In situations when the environment may change, workers need to be primed in using their judgement to inform decisions in a way that is acceptable to the organisation. PRA should be thought of as a tool to aid decision making, and central to an effective management system for health and safety, which is embedded within the organisation's business structures and processes.

If goals are clear in PRA, this will aid the decision evaluation in the field. This does not mean that you have to write absolutely everything down; PRA should not be a bureaucratic process. What we are saying is that the PRA should be sensible and proportionate to the risk. If you succinctly articulate where you are going and how to get there, with what parameters you are providing, your people will know what to do en route and are more likely to make decisions in line with your organisation's goals. However, if you set unrealistic targets, people will invariably cut corners. With PRA, don't just look at the complicated hazards; look at the simple things that can go wrong too. Remember that people overestimate perceived complex issues and underestimate perceived simple issues.

Defining hazard and risk

Chapter 2 defines a hazard as anything with the potential to cause harm; for example, working with chemicals in a laboratory, working from ladders as a window cleaner, or working with live electricity on a railway track. Linked to the hazards are the associated risks, which are concerned with both the likelihood and consequence of a specific hazard being realised. Organisations managing risk need to carefully consider the range of hazards that may arise in their work setting. There are some occupations that operate in highly hazardous environments, requiring personnel to enter dangerous situations, when most of us would be trying to get as much distance from these as possible. Even with well-managed risk systems, and with proportionate controls in place to deal with predictable risks, workers may not be able to entirely control the risk within their environment and harm may still arise.

Employer responsibilities

For the purposes of health and safety law, employers are classified as 'duty holders' and, as such, have legal obligations to manage occupational health and safety risks. Those that create occupational health and safety risks should manage them in the same way that they are expected to manage other types of risk. In the UK, the principal duty in this regard falls under the Health and Safety at Work etc. Act (HSWA) 1974. This places general duties on an employer so far as is reasonably practicable to ensure the health, safety and welfare of people at work and those affected by work activities. The specific duty to manage risk is covered under the Management of Health and Safety at Work Regulations (MHSWR) 1999. In particular, Regulation 3 requires a PRA to be undertaken covering the work. HSWA and MHSWR were referred to in Chapter 2.

PRA is a line management function and line managers have the responsibility for this within the scope of their control. To discharge this duty, managers are required to have an appropriate level of competence. We describe this as the combination of skills, knowledge, attitude, training and experience (SKATE). You'll read more about this in Chapter 7. Organisations may seek to obtain external competence (via consultants) in this area, if the competence is not available in house. However, prior to doing so, organisations should refer to the HSE website for useful health and safety tools to help them. Should an organisation engage a consultant, it should be aware that the duty to ensure that the risks are managed and the accountability for health and safety within the organisation are retained by the employer. There are separate legal obligations placed upon consultants, in order for them to demonstrate that they have the necessary competence to provide occupational health and safety services to an employer. This requirement falls under Regulation 7 of the MHSWR. In response to political pressure for competent health and safety consultants, the regulator in the UK (supported by many professional and examination bodies in the field) established an Occupational Safety and Health Consultants Register (OSHCR). The HSE strongly recommends that organisations wishing to obtain outside expertise and advice should seek such services via the OSHCR. Services are sourced via contacts on the OSHCR website at www.oshcr.org/.

Moreover, despite the huge amount of information available on PRA, its utility still remains an enigma in terms of its general under-utilisation. Part of the challenge remains that PRA is perceived as a specialist function rather than an essential part of a manager's portfolio, akin to managing and developing people, budgets and performance targets and writing business plans. The aim here is to ensure that organisations understand that risk assessment is the responsibility of management within the organisation, and not the sole responsibility of the occupational safety and health advisor.

The activity-based (predictive) assessment should be considered prior to the commencement of the activity. This will allow the manager to focus on business objectives in a safe and efficient manner, instead of having to devote additional time to dealing with urgent, predictable risks. In the predictive risk assessment, insignificant (trivial) risks can be disregarded, to allow you to focus on hazards that are significant and may cause real harm. Where they are deemed significant, the PRAs are required by law to be documented and are subject to periodic review. Likewise, where the assessment is deemed no longer valid (e.g. change of work equipment) or following an accident the PRA should be reviewed.

The aim of PRA is to create a framework for safety in the field. Managers should treat PRAs as living documents and subject to continual improvement. The PRA is undertaken in slow time in a manner that runs in line with your business operations. This allows issues to be carefully considered prior to the activity taking place. The assessment must cover all the relevant factors, e.g. all significant risks arising from the planned activity. There are requirements in law to involve and consult with the workers; this really does make sense as your workers are the ones who have experience of the day-to-day work activities. Consultation with the workforce can also include recognised trade unions or other elected representatives. It is well known that organisations that have a recognised union-based workforce perform better in occupational health and safety than those that do not.

Where organisations have similar or generic hazards, generic risk assessment is commonplace. Where the assessment has identified further actions to improve control of these risks, these should be prioritised. It is the responsibility of the employer to take necessary actions in accordance with the assessment.

Relationship of PRA to DRA

The distinction between PRA and DRA is that PRA is generally static in nature, and written down prior to the activity taking place, whereas DRA (informed by the documented PRA) is carried out in fast time, using mental processes, and provides a greater awareness of risk at a specific time in the field. This enhances the rich information in the PRA with live intelligence that can be expressed in the field. DRA is generally not recorded at the time; instead, it is generally recorded post-event as a means of learning. In some emergency responses there may be a time-stamped record of decisions made, as a part of creating the record for this learning.

The great thing about PRA is that you can competently complete this even though the activity has not yet occurred. You also have the perfect opportunity to implement the controls in order to mitigate the risks that are likely to be present. The DRA ensures that the assessment remains live during implementation. Workers need to be motivated to adopt the specified controls from the PRA and to be able to apply these in the field with adaptations where necessary. This is the dynamic piece, knowing when to modify the PRA. The aim of DRA is to support the PRA in the dynamic environment, enabling the worker in the field to be safer.

Where appropriate, the manager responsible for the PRA should make it clear that there is a need to undertake a DRA in the field. This is particularly important when the outcome could be determined by a number of variables. The worker making this DRA should retain a mental reference to the PRA.

A vexed question relating to PRA is how effectively it remains alive when it is required to be applied in the field once the task has started and the conditions begin to change.

We know that risk assessment is not an exact science and, for the purposes of DRA, any assumptions or criteria made through the PRA (where there is a level of uncertainty) should be clearly communicated to those on the ground and others that are part of the risk-based decision-making process. This should form part of the wider health and safety management system and will aid workers in the field to make informed decisions when managing risk. In doing so, workers need to feel confident and empowered to make decisions, knowing that demonstrating the required behaviours will be supported by the organisation. What the concept of DRA allows you to do is to mentally assess the situation so as to achieve the required outcome. It is correct that if an organisation's risk assessments are suitable and sufficient, there should be very few significant risks that have not previously been considered by the PRA.

Remember, the PRA should be balanced and proportionate, in terms of the level of detail against the type of operation and profile of risk. The PRA should also consider the handling of uncertainty, time-pressure issues, novel or new situations and be written in a factual and informed way. The overall aim is that risk assessment is supportive of organisational priorities, whilst taking account of the culture to mitigate against any risk. This approach will foster a positive belief about health and safety by the workforce and reduce the opportunity for poor standards to occur unchallenged.

The six steps to DRA outlined in Table 3.1 and as used by HM Fire Service Inspectorate (2002) illustrate that the approach is anchored in the PRA methodology shown in Table 3.2. Once the activity is concluded, the opportunity arises to feed back dynamic risk assessment learning from the activity or incident through debriefings (see Chapter 8).

The HSE guidance, 'Five steps to risk assessment', can be downloaded from this book's companion website (www.routledge.com/cw/dynamic-risk-assessment).

Table 3.1 Six-step approach to dynamic risk assessment (fast time)

1	Evaluate situation/activity and person(s) at risk.
2	Introduce and declare tactical mode – offensive mode (normal procedures and control measures being used), defensive mode (risks outweigh benefits, withdraw from hazard areas and operate from safe distance) or transitional mode (combination of offensive and defensive modes).
3	Select a system of work.
4	Assess the chosen system of work – is it safe? (Consider risk v benefit, if appropriate proceed).
5	If not, introduce additional controls or select another system.
6	Re-assess systems (introduce further controls if required).

Source: HM Fire Service Inspectorate (2002)

Table 3.2 Five-step approach to predictive risk assessment (slow time)

Step 1	Identify the hazards.
Step 2	Decide who might be harmed and how.
Step 3	Evaluate the risks and decide on precautions.
Step 4	Record your findings and implement them.
Step 5	Review your assessment and update if necessary.

Source: HSE (2011)

The declaration of tactical mode

In the Fire and Rescue Service, step 2 in Table 3.1 was introduced to formally compel commanders to determine if it is safe to proceed. It requires the commander to exercise judgement. This is a system applied within the Fire and Rescue Services, and other organisations may wish to consider these steps and their applicability in relation to the specific requirements of their own organisations.

The role of human factors

One of the important areas to consider with all types of risk assessment is the human factors element. The study of human factors from a risk perspective is a complete subject in its own right. When seeking to understand a worker's behaviour, human factors should be considered as part of the PRA. Gutteling and Wiegman (1996) suggest that people are likely to respond to risk in the context of their own experiences, and these contexts will vary with each individual's own experience. This probably explains why people do not always evacuate at the sound of an alarm – fire doesn't happen here.

Line managers who have responsibility for their workers in the field need to recognise their own biases when undertaking PRAs. It is important to be aware of biases so that we are less likely to be affected by them. In the field, this may be magnified if you have little or no confidence in the PRA and/or the environmental conditions are stressful and other factors will affect your ability to make effective risk judgements. You should also be mindful of cognitive dissonance. This is concerned with an individual's behaviour where it conflicts with their belief system. An example of this is where an individual is aware that smoking is harmful to their health, but continues to smoke. You should try to ensure that workers' values and beliefs are consistent with the organisation's values. If employees genuinely buy in to the values of the organisation they are more likely to abide by them.

There is an increase in research on human behaviour and the implications for occupational safety and health. Our understanding of the significant role behaviour plays in accident causation at work is informed by the HSE, which reports that people's actions or omissions have contributed in some part to an estimated 80 per cent of accidents (HSE, 2009a).

Human failures impact across the organisation at all levels. Every human being makes mistakes, which may cause or contribute to an accident, regardless of individual experience and level of training. Major incidents have arisen which unfortunately bear witness to this. The *Piper Alpha* disaster (examined in Chapter 6) identifies the failings of people at all organisational levels and led to 166 fatalities. Equally, individuals can prevent accidents by their own positive actions. Risk assessment is the cornerstone of accident prevention, and the HSE publication *Reducing Error and Influencing Behaviour* (HSE, 2009a) – available for download from this book's companion website (www.routledge.com/cw/dynamic-risk-assessment) – explains how and where failures can arise and provides useful insights into how these can be mitigated.

It is always important to consider cognitive biases that can have an effect on decisions. An example of this is 'risk compensation' (see Table 1.1). This is the phenomenon where additional controls have been put in place to protect workers, but these controls give a false sense of added security, resulting in workers taking greater risks; for example, in snowy and icy weather using 4x4 vehicles, instead of vans, to deliver goods to vulnerable clients. The driver may take greater risks than they would normally, as they believe that the 4x4 can tackle almost any environment. Another example could be the issuing of stab-proof vests, where the wearer feels protected and therefore ignores or overlooks safety procedures and tackles an individual single handed. People can also develop the fallible view of illusionary optimism, 'it will work out in the end' (see Table 1.1). They may have a tendency to concentrate on positive information when making decisions about themselves. The only situation where this is not the case is in people who suffer from depression; they tend to focus on negative information.

The theory of reasoned action (TRA) proposes that in predicting how an individual will behave, we have to look at behavioural intention. This is argued to be determined by the individual's attitude towards carrying out the particular behaviour (e.g. to jump into a river to save a child's life) and the subjective norms (the perceived social pressure or expectations of a valued group or person) (Fishbein & Ajzen, 1975). Another aspect is that an individual's attitude is made up of the behavioural belief about whether the outcome of the behaviour will be positive or negative and their belief about what other people (valued people or groups) may feel about the behaviour (normative beliefs).

A person adopts a particular behaviour because it accords with their existing beliefs or values. This is called 'internalisation'. The link between attitude and behaviour at this point is stronger, as the individual is engaging in this behaviour because they believe it is right, regardless of external influences. Therefore, if a police officer feels that they have a personal duty to place themselves in danger in order to protect the public or colleagues, they are more likely to violate police standard operating procedures and place themselves in danger. This is also more likely if they believe that people they respect would expect them to do this (e.g. their sergeants and colleagues). Ajzen (1991) recognised limitations to the TRA theory, in that it did not account for situations where people feel they have little or no control over their behaviour. The concept of perceived behavioural control refers to the individual's belief in their own ability to complete a task (how easy or difficult a task/behaviour will be) and the amount of control they have over the attainment of the task. Ajzen (1991) added this concept to the original TRA model and this became the theory of planned behaviour. The authors suggest that organisations need to be primed to tackle these issues and to seek a wider understanding of cognitive theories when developing policies, risk assessment, safe systems of work and operational training.

Policing challenges when facing risk

Within the Police Service there is a common purpose amongst sworn officers to carry out operational duties and, where necessary, to operate in environments where significant risk may be present, known as the 'policing imperative' (Waters, 2001).

McBride (1996: 5) reported that the common purpose of the Police Service is to:

> uphold the law fairly and firmly; to prevent crime, to pursue and bring to justice those who break the law; to keep the Queen's peace, to protect, help and reassure the community; and to be seen to do this with integrity, common sense and sound judgement.

The Association of Chief Police Officers' (ACPO, 2011) decision-making model is encapsulated within its statement of mission and values at the heart of its decision-making process:

1 Gather information and intelligence.
2 Assess threat and risk and develop a working strategy.
3 Consider powers and policy.
4 Identify options and contingencies.
5 Take action and review what happened.

> The mission of the police is to make communities safer by upholding the law fairly and firmly; preventing crime and antisocial behaviour; keeping the peace; protecting and reassuring communities; investigating crime and bringing offenders to justice.
>
> (McBride, 1996: 5)

There is a public expectation that a police officer will assist whenever called upon; however, in certain situations this expectation may be problematic. The 'first duty of a police officer is to preserve life' and the 'overall mission of policing is public safety' McBride (1996: 5). Each operational incident will vary and can be complex and emotive, and there may be a conflict between protecting life, reducing crime and controlling the risks for police officers. A situation where this may arise is where there are personal safety issues involved. For example, a highly hazardous situation could entail a child struggling to remain afloat in a fast-flowing river and a police officer is the first on the scene (without specialist equipment or back-up). The officer is likely to be pressured by a distressed relative or member of the public who may have raised the alarm to attempt a rescue. In this type of emergency situation the officer may have little or no experience of dealing with a water rescue (Stewart, 1997).

This type of situation has the characteristics of time pressure and inherent dangers where the officer has to make a safe decision very quickly (Brehmer, 1992). Research by Ritchie and Marshall (1993) and Kemp et al. (1992) has found that decisions are taken that appear most likely to satisfy the organisation's goals or objectives. These organisational goals may not always be understood or in harmony with public expectations. Local and organisational policing goals play a key role in the decision-making choices available to police officers when they deal with routine or difficult situations. These would include values and beliefs held by the individual, colleagues and management (Kemp et al., 1992). Cognitive theories play a key role, however, and Fielding (1994) reported that whilst a police officer will recognise the present dangers they may feel strong social pressure to attempt a rescue, as they believe that this is expected of them. They may also feel that they are expected by fellow officers to attempt a rescue in order to prove that they are tough. This is sometimes known as 'canteen culture'. Sterling (1972) comments on how quickly new police recruits adopt these norms.

Case study 3.1   The Hillsborough disaster

This case study takes you through some of the events, with reflections by Mrs Margaret Aspinall, who, like all the relatives and friends of the bereaved and injured, has been profoundly affected by the events of 15 April 1989.

The Hillsborough disaster is a poignant reminder of how overlooking health and safety can have a devastating effect on many lives. The search for answers and the steadfast resolve of the victims' families and friends have highlighted the catastrophic effect of poor health and safety management in this case.

Margaret's son James Gary Aspinall was an 18-year-old Liverpool supporter. This was his first away football game. It was the FA Cup semi-final between Liverpool and Nottingham Forest, held at Hillsborough stadium, home of Sheffield Wednesday Football Club. Ironically, this game was a repeat of the FA Cup semi-final between these two teams that was held the year before without incident. Over 50,000 men, women and children travelled to the game and James was one of the 24,000 Liverpool supporters. It was a fine spring day, and James arrived at the Hillsborough stadium via coach at approximately 13.20 hours. James and other Liverpool fans were to be situated in the north end of the stadium at Leppings Lane, which consisted of several pens. It is reported that the supporters entered the pens through an unsigned tunnel which had a perimeter fence and a locked gate at the front of each pen.

It was reported that, due to the number of people trying to enter through the antiquated turnstiles, there was unavoidable pushing and jostling down the tunnel, prior to supporters getting into the pens. James was in pen 3 and it became apparent that the standing-only pens 3 and 4 had very quickly become overcrowded, whilst it was reported that other pens had not been filled. The pen 3 safety barrier broke, due to the sheer volume of people, and it took a while before the seriousness of the overcrowding was realised. Just before kick-off at 14.52, there were vast crowds of several thousand people still outside the ground trying to get into the stadium. A senior officer gave the order for gate C to be opened. This resulted in thousands of supporters getting into the ground without being directed, and most headed for the central tunnel leading to the already overcrowded and confined pens 3 and 4. This overcrowding greatly exceeded the safe capacity and led to crushing. Many supporters became trapped, whilst others attempted to move into pen 4 to try to escape the crush. By this time, it was highlighted that many people had collapsed, due to the intensity of the crush. The match had kicked off at the designated time of 15.00 and some of the horror on the terraces as it unfolded was captured on the televised game, which was being watched by millions, including many of the family and friends of the fans at the game. Supporters eventually spilled onto the pitch to escape the crush and the match was stopped after six minutes, at 15.06.

As a result of the crush and overcrowding, 96 men, women and children died and several hundred were injured, with many more being traumatised by what they had experienced. James was one of the 96 supporters who never came home. This became known as the worst football disaster of our time. As a result of the tragedy, the majority of the family and friends who lost loved ones at Hillsborough Stadium, Sheffield founded the Hillsborough Family Support Group in May 1989 in their memory. Figure 3.1 shows the Hillsborough memorial erected at Liverpool Football Club.

Issues highlighted by the Hillsborough Independent Panel Report (2012)

* Previously known incidents associated with this type of sporting event, e.g. at Hillsborough stadium in 1981:
  'Semi-final between Tottenham Hotspur and Wolverhampton Wanderers, there was serious congestion at Leppings Lane and crushing on the confined outer concourse', led to many injuries and a slender avoidance of fatalities.

* Crowd management:
  'It is evident from the disclosed documents that South Yorkshire Police were preoccupied with crowd management, segregation and regulation to prevent potential disorder. Sheffield Wednesday Football Club's primary concern was to limit costs'. 'The Fire Service, however, raised concerns about provision for emergency evacuation of the terraces.' 'As the only means of escaping forwards was onto the pitch, concern was raised specifically about the width of the perimeter fence gates which was well below the standard recommended by the Green Guide.'

* Outdated Safety Certificate:
  'Following the near tragedy in 1981, Hillsborough was not used for FA Cup semi-finals until 1987. During this period the Leppings Lane terrace underwent a series of significant modifications and alterations, none of which led to a revised safety certificate.'

Figure 3.1 Hillsborough Memorial
Source: Photograph provided by Liverpool Football Club

- The condition and adequacy of the turnstiles and the co-ordination of access to the central pens via the tunnel:
  'Recommendations to feed fans directly from designated turnstiles into each pen, thus monitoring precisely the distribution of fans between the pens, were not acted on because of anticipated costs to Sheffield Wednesday Football Club.' 'Consequently, turnstile counters were rendered irrelevant. Although they provided a check on the overall numbers entering the terrace, there was no information regarding crowd distribution between pens, each of which had an established maximum capacity.'

- Capacity of the terrace, crowd monitoring of capacity within the pens:
  '[M]aximum capacity for the terrace was significantly exceeded' and a 'lack of precise monitoring of crowd capacity within the pens'; 'safety of the crowd compromised at every level'.

- Replacement of an experienced match commander 21 days prior to kick-off:
  'The challenges and responsibilities of policing and managing capacity crowds at Hillsborough were evident following the events of 1981 and the subsequent difficult relations between South West Yorkshire Police (SYP) and Sheffield Wednesday Football Club. In this context, the decision by SYP senior management to replace an experienced match commander just 21 days before the match is without explanation in the disclosed documents.'

- Emergency egress/design issues:
  Deficiencies in: 'alterations to the terrace, particularly the construction of pens; the condition and placement of crush barriers; access to the central pens via a tunnel descending at a 1 in 6 gradient; emergency egress from the pens via small gates in the perimeter fence'.

- Inadequate debriefs from previous FA Cup semi-final games:
  'Crucial information arising from these events was not shared within South West, nor was it exchanged between South West Police and other agencies. There is no record provided by Sheffield Wednesday Football Club of debriefings held between Club stewards and their managers.'

- Emergency plan response:
  'Communications between all emergency services were imprecise and inappropriately worded leading to delay, misunderstanding and a failure to deploy officers to take control and coordinate the emergency response.'

- Location of Liverpool supporters in the terraces and pens in the smaller end of the ground:
  'The confined outer concourse area serving the Leppings Lane turnstiles accommodated the entire Liverpool crowd, heading towards three discrete areas within the stadium (North Stand; West Stand; Leppings Lane terrace). It was a well-documented bottleneck and at matches with capacity attendance presented a predictable and foreseeable risk of crushing and injury.'

The Hillsborough Independent Panel Report highlighted multiple factors which led to the tragedy of 96 deaths, none of which was attributable to the supporters.

It is evident that DRA as we express it in this book would not have been applicable in this situation, due to shortcomings or absence of strategic and predictive assessment.

A copy of the independent Hillsborough panel report is available on this book's companion website (www.routledge.com/cw/dynamic-risk-assessment).

TIP

Managers and those who are responsible for activities should ensure that predictive risk assessments are underpinned by safe systems of work.

Example: An outdoors instructor organised a swimming event that involved each participant having to dive into the water. One of the participants couldn't swim and had to be rescued from the water.

Solution: When organising events that require a level of skill, make sure the participants are able to fulfil any skill requirements. For example, have a system in place that requires instructors to test swimming ability in the shallow end first.

TIP

Organisations seeking desired behaviours from its workforce should align its expected behaviours and core values to risk management systems, in the same way it aligns service delivery expectations.

Example: A refuse and recycling operative responsible for collecting waste from street properties is struck and seriously injured by an oncoming vehicle during the early hours of a damp and foggy morning. The operative had hastily descended from his refuse truck in order to collect a forgotten refuse bag from the opposite side of the road. He was not wearing his hi-visibility jacket.

Solution: If senior management lead by example by following the organisation's policies and procedures, then staff are more likely to follow. Management should take appropriate action against staff who fail to adhere to established safe systems of work. Dependant on the nature of the incident, a range of actions could be considered such as: training/re-training, shadowing from a more experienced member of staff, verbal warnings or a range of disciplinary options.

TIP

Major public-facing organisations need to ensure that their workforce is suitably prepared to deal with and manage public expectations when dealing with risk.

Example: A member of the security staff working for a Local Authority witnessed a man entering the reception area with what appeared to be a weapon. He was shouting abuse at anyone that caught his eye and then pushed an elderly lady to the floor. A bystander noticed the surprised security guard and pleaded with him to tackle the man.

Solution: Employers have a responsibility to impress upon their workforce their expected behaviours and role. This type of situation is predictable (for the activity of a security guard role) and as such, staff should be given a clear understanding of what the role requires (including commensurate training) when dealing with situations of this nature.

TIP

Organisations need to consider how groups make risk-based decisions to ensure effective decision making.

Example: An employee raises a safety concern affecting a production line that requires technical or specialist knowledge. This is brought to the attention of senior management. Several managers discuss the issue and decide to follow the general consensus of opinion, which is to continue the production line (during a busy period), as they believe the concerns raised do not pose a serious risk to workers. In reaching this decision they disregard the strongly held concerns of one manager in the meeting.

Solution: It is too easy for a dissenting voice (who may hold a valid concern) to be ignored by a group. Members of a group may agree with the majority of managers or withhold vital information for a number of reasons, such as to maintain group harmony or because they doubt their own opinion (with so many others holding a different belief). Different approaches to gathering information/evidence should be considered to avoid these types of group biases. For example, each person should document their thoughts/opinions in isolation (prior to discussion) to avoid the influence of other group member opinions.

Chapter summary

Before you move on to the next chapter, here is an opportunity for you to check your knowledge.

1 Why is it important for organisations that procure the services of external consultants to have the appropriate level of occupational safety and health competence?
2 Why would the application of DRA (in isolation) not be applicable to the Hillsborough tragedy?
3 Why is it important to recognise your biases when undertaking PRA?
4 What are the benefits of obtaining information from employees about the hazards when undertaking PRA?
5 How is PRA defined?
6 At what stage would you undertake a PRA, and why?
7 How can understanding human factors in relation to worker behaviour improve health and safety performance?
8 Why is PRA vital to the application of DRA?
9 What is risk compensation?
10 What is the main difference between PRA and DRA?

Chapter 4

The 3-Level Risk Management Model™

Many incidents are characterized by time pressure, where the commander has to make decisions very quickly in order to save life and to prevent the escalation of the problem. This appears to be regarded as a stress factor for commanders and their teams, and may be particularly acute in the opening stages of an incident. Brunacina (1985) advises commanders that 'most of the time on the fireground the first five minutes are worth the next five hours'.

Flin (1996: 110)

[E]ven when all reasonably practicable precautions have been taken to deal with foreseeable risks, injuries and deaths could still occur; and it may be necessary to take some risks to secure a wider benefit to public safety.

HSE (2009)

You have acquitted Greater Manchester Fire and Civil Defence Authority. They have been found not guilty of any breach. But for the families of Reyaz Ali and Paul Metcalf what occurred on September 5 1999 was a personal tragedy which no doubt still lives on for them. I would like to express my own sympathies to them.

Judge William Morris (Morris, 2013)

Introduction

This chapter introduces and explains The 3-Level Risk Management Model. The model provides the most evolved and up-to-date expression of risk assessment thinking available in the world today, yet it is built upon a researched evolution from a series of grounded theories established since 1998.

Our model highlights three levels for making decisions and assessing risks in the organisation:

1   strategic level
2   predictive level
3   dynamic level.

We provide guidance on how this model can be adapted, implemented and integrated into the structure and the discipline for decision making at all levels in your organisation. For some, this may be a small adaptation, as the model builds upon and complements many common management approaches, including (strategic) business planning and meeting legal requirements. We also provide exemplar case studies and other examples in this and each of our other chapters. We explain how the continuous loop of improvement in risk assessment and risk-based decision making can be achieved.

The 3-Level Risk Management Model™

In developing our model, we have taken account of the established risk assessment, dynamic risk assessment (DRA) and decision-making models used by other organisations, and as exemplified throughout this chapter and book.

Whilst we are informed by those before us, we have noticed opportunities to strengthen and reinforce prior learning, and this is incorporated. We have also learned from the interviews, case studies and other examples in this and each of our other chapters.

Our model is shown in Figure 4.1 and highlights the organisational context, the identification of business risk, and the three levels for assessing those risks and making decisions: the strategic level, the predictive level and the dynamic level. It provides for a consistency of approach at all levels in the organisation.

Context

Your organisation is not operating on the Moon. There are pertinent features of geography, a legal framework, competitors and partners, financial freedoms or constraints, and the status of technological development within your operating territory to take account of. These represent just a tiny sample of possibly significant factors arising in your business context. A 'PEST' [political/legal; economic/macro-financial; social/demographic; technical/infrastructure] analysis (or similar) will assist an organisation to identify a fuller inventory.

Understanding this context for your organisation is fundamental to identifying and addressing risks at any level, as some arise in the environment. The International Standards Organisation (ISO) published *Annex SL* in April 2012. It is anticipated that this will have a significant impact for all ISO management systems (ISO, 2012), as it requires all organisations recognising the ISO methodology to take account of the context of the business environment. It also represents best practice for non-certified organisations, and we recommend it to you.

All ISO technical work, including the development of standards, is carried out under the overall management of the Technical Management Board

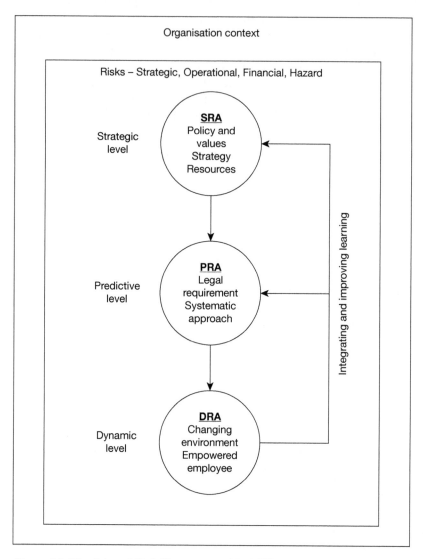

Figure 4.1 The 3-Level Risk Management Model™

(TMB). ISO/TMB produced *Annex SL* with the objective of delivering consistent and compatible management system standards (MSS). The Annex describes the framework for generic MSS and requires the addition of discipline-specific requirements to make fully functional standards for systems such as quality, environmental, service management, food safety, business continuity, information security and energy management.

In the future, all new MSS will have the same overall 'look and feel', thanks to *Annex SL*, and current MSSs will migrate during their next revision. This should be completed within the next few years. At the time of writing, ISO 14001 is in the final stages of redevelopment, and ISO 9001 and BS OHSAS 18001 will undoubtedly follow.

Risks

ISO 31000 defines 'risk' as the effect of uncertainty upon objectives or, to put it even more simply, the impact upon objectives (Asbury, 2013). Some risks will arise from outside the organisation (such as adverse weather) and some from inside (such as equipment and the quality of training provided).

The Institute for Risk Management (IRM, 2002) identifies risks arising externally (E) or internally (I) in four families:

1 Strategic risks – e.g. demand and competition (E), merger and acquisition (E/I), intellectual capital (I)
2 Operational risks – e.g. regulation (E), recruitment (E/I), accounting (I)
3 Financial risks – e.g. interest rates (E), liquidity and cash flow (I)
4 Hazard risks – e.g. natural events (E), employees [health and safety] (I).

Risks can be identified in a variety of ways, but most often we notice that these are disconnected. Our model connects these three levels of assessment, and provides the roadmap (and the mind map) for the connectivity between these risks and for driving integration and improvement.

Three levels

1 Strategic level: Risk assessments concern policy and strategy decisions, developing the culture and reputation, provision of high-level plans and resources, acquisitions and divestments, purchasing. These decisions are made, and informed over time, by the results of predictive-level assessments.
2 Predictive level: Risk assessments concern the documented arrangements to meet legal requirements, including MHSWR. These operational assessments are made by line managers within the constraints of the strategic-level decisions of current or future tasks, roles and activities, leading to prioritisation using a hierarchy for improved control. These decisions are made and informed over time by feedback from dynamic-level assessments.
3 Dynamic level: Risk assessments concern real-time decisions, which rely upon the resources provided at the strategic level to provide structured control at the predictive level, and are then adapted to take account of

conditions on the ground. Workers are empowered to make decisions, but are expected to be able to explain them. Capturing learning after the event allows feedback and learning to the strategic level and the predictive level. The aim is 'better next time'.

Time and pace

The time available, and the pace required at different levels, varies and concerns the amount of time available to those involved in which to decide.

Strategic-level decisions require thought, research, a quest for funding and other long-term matters, and thus may take months or years to conclude. We also recognise that some strategic decisions may require a very quick call (e.g., 'go for launch').

Predictive-level assessments have greater urgency, as there is often a legal requirement to assess risks (particularly those relating to health and safety). A regulator's improvement notice may also condense the time available. However, in most cases, a predictive risk assessment will take weeks (or months) to consult, prepare and communicate. Also, additional time may be required to inform, instruct, train and supervise the control measures in use.

Dynamic assessments have the greatest urgency, and may take seconds or minutes (or hours) to decide the course of action.

Integrate and improve

The 3-Level Risk Management Model™ highlights the learning loop of integrating good ideas and learning from past events into the predictive- and strategic-level assessments for the future. This inevitably leads to learning about past events and how those decisions leading to success or failure were made.

Chapter 5 of this book provides guidance on how to embed and integrate The 3-Level Risk Management Model™ into the structure and discipline for decision making in your own organisation.

Case study 4.1 NASA: 'The hot seat'

An Act to provide for research into the problems of flight within and outside the Earth's atmosphere, and for other purposes.

(National Aeronautics and Space Act, 1958)

With this simple preamble, the Congress and the President of the United States created the National Aeronautics and Space Administration on 1 October

1958. NASA's birth was directly related to the pressures of national defence. After the Second World War, the US and the Soviet Union were engaged in the Cold War, a broad contest over the ideologies and allegiances of the nonaligned nations. During this period, space exploration emerged as a major area of contest which became known as the space race (NASA, 2013).

In over sixty-five years, NASA has launched 166 manned missions in its Mercury, Gemini, Apollo and Space Shuttle programmes, taking 370 astronauts into space, and 12 to the surface of the Moon. Twelve other astronauts circulated the Moon without landing. There have been 164 safe landings; seven astronauts were lost 73 seconds after the launch of Shuttle flight STS51L in 1986, and seven more were lost 15 minutes before their return to Earth upon the disintegration of Shuttle STS107 in 2003. You can read more about Space Shuttle *Columbia* (STS107) on pages 120–121.

NASA has also dealt with other serious incidents, involving a launch-pad fire in 1967 (Apollo 1) and an oxygen tank rupture and its aftermath aboard the Moon-bound Apollo 13 (1970). Although this mission never landed on the Moon, it reinforced the notion that NASA had a remarkable ability to adapt to unforeseen technical difficulties inherent in human spaceflight. How does NASA achieve this?

Launch control – Kennedy Space Center, Florida

All manned NASA launches have been conducted from Kennedy Space Center (KSC) since Apollo 8 on 21 December 1968. The facility incorporates the launch control center (LCC), the vehicle assembly building and offices for telemetry, tracking and instrumentation equipment, the launch system and four firing rooms.

On launch days, there are typically a dozen key specialist personnel, headed by the Launch Director (LD). The LD is the head of the launch team and is responsible for making the final 'go' or 'no go' decision for launch after polling the team members. Any member of the team can abort the launch, but only the LD can authorise it.

Mission Control – Johnson Space Center, Texas

The responsibility for spacecraft launches remains with the LCC until the rocket has cleared the launch tower. Then, responsibility is handed over to NASA's Mission Control Center (MCC) right through until landing days, sometimes weeks later. MCC Houston is shown in Figure 4.2.

Figure 4.2 NASA Mission Control

MCC, Houston comprises two Mission Operation Control Rooms (MOCR, pronounced 'moh-ker'): MOCR1 and MOCR2. These are latterly called Flight Control Rooms (FCR, pronounced 'ficker'). NASA controlled its first four missions from MOCR2 (shown in Figure 4.3), before transferring flight operations to MOCR1.

From the MOCR, the Flight Director (FLIGHT) monitors the activities of a team of twelve to seventeen flight controllers, each responsible for a particular element of the mission. The Flight Controllers' Creed (NASA, 2013) says that they must 'always be aware that suddenly and unexpectedly we may find ourselves in a role where our performance has ultimate consequences'.

FLIGHT consults with the flight controllers, but holds the overall operational responsibility for mission and payload operations and for all decisions regarding safe, expedient flight. FLIGHT can overrule the US President on mission decisions (NASA, 2013).

The capsule communicator (CAPCOM) is the only member of the MCC team to communicate directly with the crew of a manned space flight. The role of CAPCOM is filled by an astronaut, often one of the back-up or support-crew members. NASA believes that an astronaut is most able to understand the situation in the spacecraft and pass information in the clearest way.

Figure 4.3 MOCR2

It was in MOCR2 that the MCC team practically applied the two stages in the process of making a decision discussed later in this chapter to successfully resolve the incident aboard Apollo 13.

NASA remains at the forefront of technical development and, in collaboration with General Motors and Oceaneering, has designed a state-of-the-art, highly dexterous, humanoid robot called Robonaut 2 (R2). R2 is made up of multiple component technologies and systems including vision systems, image recognition systems, sensor integrations, tendon hands, control algorithms and

Figure 4.4 Robonaut 2

much more. R2's nearly fifty patented and patent-pending technologies have the potential to be game-changers in multiple industries, including logistics and distribution, medical and industrial robotics, and beyond.

During a recent visit to the Johnson Space Center, we were treated to meeting R2 (shown in Figure 4.4). Robonauts have already flown into space. R2 was launched to the International Space Station on space shuttle *Discovery* as part of the STS-133 mission. It was the first dexterous humanoid robot in space, and the first US-built robot at the space station. But that was just one small step for a robot, and one giant leap for robot-kind.

Authors' comment: strategic–predictive–dynamic

We observe the following characteristics in relation to our risk management model shown in this chapter.

1 Strategic: NASA has followed its founding principles for over sixty-five years and, despite several tragedies, has stuck to its purpose with a general expectation of mission success. Strategic decisions to eliminate

risks to human astronauts have led to the development of robots for routine tasks on missions to space.

2   Predictive: Based on detailed mission assessments, and the decisions of specialists, operational decisions are communicated in 'astronaut language' by one individual (CAPCOM).

3   Dynamic: Decision making is by specialists and subject matter experts (quite literally 'rocket scientists'), who can abort launches or operations in space. Everyone is involved. 'Go' decisions are centred on the FD, who is very powerful and can overrule the executive if necessary. The diagnosis of the problems on Apollo 13, the creative decision making and the clear communication from CAPCOM saved the mission, and saved the crew of three astronauts from certain death.

Other risk management models, 1998–date

Fire and Rescue Services Operational Risk Management (ORM)

As noted in Chapter 2, the ministers with responsibility for the Fire and Rescue Service (FRS) of the UK endorsed an approach to operational risk assessment in July 1998 (HM Fire Service Inspectorate, 1998). The approach focused upon meeting the moral, economic and legal reasons for the FRS to take the management of health and safety seriously. It also enabled the service to successfully deal with the HSE improvement notices that had been served upon it since the deaths of several fire fighters. Table 4.1 shows the number of operational and other fire-fighter deaths in the period 1978–2008: a total of 122 (82 operational and 40 others). These include sixty-five whole-time fire-fighters, thirty-three retained and the balance (surprisingly) made up of unknown/undeclared employment status. Operational deaths relate to those occurring in circumstances where the individual was either en-route to, at, or returning from a fire, water or other incident. Other deaths include non-operational road traffic incidents and deaths during training.

Early success for ORM

For seven consecutive years following the introduction of ORM, the three-year rolling average number of operational fire-fighter deaths remained below two per year, from a high of four per year in the early 1980s, suggesting that the new approach was probably successful. However, specific events in the Fire and Rescue Service since (see the example in our case study later in this chapter) serve to further highlight this priority.

Table 4.1 Fire-fighter fatalities, total and at operational incidents in the UK, 1978–2008

	Total	Operational	Three-year rolling average (operational)
1978	8	4	
1979	2	2	
1980	3	3	3.00
1981	6	5	3.33
1982	6	4	4.00
1983	6	3	4.00
1984	4	4	3.67
1985	7	4	3.67
1986	2	0	2.67
1987	7	7	3.67
1988	1	0	2.33
1989	5	1	2.67
1990	6	3	1.33
1991	4	3	2.33
1992	4	3	3.00
1993	4	4	3.33
1994	1	1	2.67
1995	3	3	2.67
1996	6	5	3.00
1997	2	0	2.67
1998	2	1	2.00
1999	2	2	1.00
2000	6	1	1.33
2001	2	0	1.00
2002	2	0	0.33
2003	1	1	0.33
2004	5	2	1.00
2005	2	2	1.67
2006	3	3	2.33
2007	8	7	4.00
2008	1	1	3.67

Source: FBU (2008)

Levels of ORM

The Fire Service ORM model provides for three levels of operational risk management.

1 Brigade command and Fire Authority provide policy, priorities, resources and a positive health and safety culture.
2 Brigade departments assess generic risks and make recommendations to improve health and safety, including for the development and implementation of additional control measures.

3   Operational personnel continuously evaluate and manage risks at incidents. After the event, incidents are debriefed.

Dynamic management of risk concerns the continual assessment and implementation of actions to control risk at an often rapidly changing operational incident. Fire Services give the responsibility for this to the incident commander. Whilst the commander can make decisions that, so far as is reasonably practicable, eliminate or reduce risks to acceptable levels, individual fire-fighters must also be aware of their own and their colleagues' safety, as they may be working in autonomous teams. All ranks have authority and a duty to take action in the interests of safety – this is an important element of the 'Safe Person Concept', described in Chapter 5.

Policing

Gold, Silver, Bronze

The terminology Gold, Silver and Bronze was developed by the emergency services to indicate the level of command and decision-making authority. Some services use 'Strategic', 'Tactical' and 'Operational' designations, which mirror Gold, Silver and Bronze, respectively.

Officer safety system

Policing is often different to operating as a fire service officer; police officers often work alone. McBride (1998) says that assessing operational police incidents is critical to officer safety. Officer safety as a theme is dynamic as new equipment and techniques are introduced to counter new threats, and an Officer Safety System is recommended for all Police Services. Such a system is shown in Figure 4.5.

The objective of the system is to improve safety by reducing the dangers that officers face. As shown, the inputs to the process are:

- personnel – more officers can be provided
- training – new training methods can be initiated
- equipment – better equipment can be procured
- procedures – new or changed procedures can be implemented.

The measurable outputs shown (and there are possibly others) are:

- morale
- assaults on police officers
- sickness and injury
- use of force.

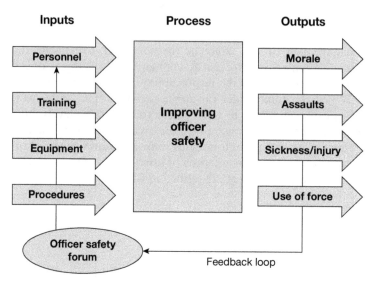

Figure 4.5 Officer safety system
Source: Abridged from McBride (1998: 181)

Standards must be set on the outputs. For example, if 90% of incidents involving handcuffing resulted in successful, injury-free arrests, this may mean that the use of handcuffs is effective. If only 10% were successful, the input(s) should be analysed and improvements initiated. The officer safety forum is established to compare the outputs with the objective of improving officer safety.

Streetcraft

Streetcraft is a common name given by some UK police forces to their officer safety programmes. Active officers typically attend streetcraft training during their probationary period and at annual intervals thereafter.

The content of streetcraft training develops year on year in the light of experiences shared within each Police Service, and through sharing between services. Whilst streetcraft training is a taught class, it is reinforced on the job:

> Some of that comes through training but a lot of it comes through learning on the job and sergeants play an important part in the role of teaching young or newly qualified PCs how you do policing in practice and how you do it well.
>
> (Jon Collins, Deputy Director, Police Foundation
> in Metropolitan Police Federation, 2013)

National Decision Model (NDM)

The UK police National Decision Model (NDM; ACPO, 2011) was introduced in 2012 to replace its predecessor, the conflict management model (commonly known as 'CMM'). In reality, there is not much difference between them. Its authors, the Association of Chief Police Officers, felt that the CMM provided an excellent approach to structure in decision making, but that its name discouraged use.

The NDM has five stages each of which is a step in the decision making process. It is a cyclical model, thus allowing decisions to be reassessed, improved and debriefed in light of new information.

A briefing note on the NDM can be downloaded from this book's companion website (www.routledge.com/cw/dynamic-risk-assessment).

Aviation: Crew resource management (CRM)

Crew (or cockpit, or flight deck) resource management comprises a set of training procedures used in workplaces where human errors can result in catastrophic consequences. CRM was originally developed for improving aviation safety by focusing on the quality of leadership, interpersonal communication and decision making in cockpits.

CRM originated as an NTSB (US National Transportation Safety Board) recommendation following an investigation into an air crash in Portland in 1978, where the captain repeatedly ignored information that his aircraft was low on fuel.

The NTSB landmark recommendation said that it would:

> Issue an operations bulletin to all air carrier operations inspectors directing them to urge their assigned operators to ensure that their flight crews are indoctrinated in principles of flight deck resource management, with particular emphasis on the merits of participative management for captains and assertiveness training for other cockpit crewmembers.
>
> (Federal Aviation Administration, 1978)

United Airlines was the first airline in the world to provide CRM training to its aircrews, from 1981. 'Sully' Sullenberger was the facilitator of the CRM course for US Airways (Sullenberger, 2009). He tells of how learning from United Airlines flight 232 taught him a great deal about flying:

> On 19 July 1989, a DC10 aircraft was travelling from Denver to Chicago with 296 passengers and crew on board. At 37,000 feet, the center engine high on the tail exploded. It caused the hydraulics for all three engines to fail. The captain discovered that he could turn right but not left. It could not land in this configuration. An off-duty pilot travelling as a passenger offered to help. The captain welcomed him into the cockpit. Sullenberger

says that this type of emergency was so rare there was no training, and no checklist to follow. [The later aviation incident enquiry reported that the chance of a simultaneous failure of these three hydraulic systems was a billion-to-one (10^{-9})] The cockpit team had to find a new way to fly the plane. Traditionally in the airline industry, there is a steep hierarchy in cockpits. Many first officers were reluctant to make suggestions to captains. That Captain Haynes welcomed input that day helped the crew to solve the problem and have a better chance of landing safely. The aircraft landed at Sioux City airport, but when the right wing struck the runway, causing it to tumble, there were 111 fatalities, yet 185 survived that day because of the work of Haynes and his crew.

(Sullenberger, 2009)

In CRM training, US Airways flight 232 is considered one of the best examples of a captain leading a team effort, while being ultimately accountable for decisions and the outcome.

Since 1990, CRM training has been modified to be used in other activities, including air traffic control, ship handling, fire-fighting and medical operating rooms, where people make high-risk, time-critical decisions.

Institute of Risk Management (IRM)

The IRM says (IRM, 2002) that risk management is a rapidly developing discipline and there are many and varied views and descriptions of what risk management involves, how it should be conducted and what it is for. It says that risk management is not just something for corporations or public organisations, but is for any activity, whether short- or long-term.

The IRM model aligns risk to the achievement of business objectives and indicates that, once risks are assessed and reported, this should lead to decision making and risk treatment. Risk treatment is the process of selecting and implementing measures to modify the risk. This implicitly includes making dynamic, fast-time decisions in the field.

The IRM says that there are many ways of achieving the objectives of risk management and, in producing its standard (which is available for download from this book's companion website – www.routledge.com/cw/dynamic-risk-assessment), it was never intended to produce a prescriptive standard that would have led to a box-ticking approach, nor to establish a certifiable process. It says that the standard represents best practice against which organisations can measure themselves.

ISO 31000:2009

ISO 31000 *Risk Management – Principles and Guidelines* was published in 2009. It provides generic principles, a framework and process for managing risk. It can be used by any organisation regardless of its size, activity or sector. Its authors say that using ISO 31000 can help organisations to increase the likelihood of achieving objectives, improve the identification of opportunities and threats and effectively allocate and use resources for risk treatment. As noted above, this implicitly includes dynamic assessments, decisions and actions in the field.

The IRM guidance on enterprise risk management (IRM, 2009) and ISO 31000:2009 (ISO, 2009) are available for download from this book's companion website www.routledge.com/cw/dynamic-risk-assessment.

Decision making

Flin et al. (2008) say there are two stages in the process of making a decision:

1   What is the problem? and
2   What shall I do?

What is the problem?

Situational assessment is critical for decision making. Fire commanders call this 'size-up'. Experienced workers grow to know what 'good' looks like, and retain memories of this state. When observation or other senses detect a change in this state, more focused situation awareness takes place. It's like driving a car: when a dashboard light comes on, that cue leads the driver to focus on the specifics of that signal (e.g. loss of tyre pressure).

Incident commanders need good situational awareness. The ability to notice the cue, evaluate and focus on a few key signals from an unfolding situation is key to success in the role.

Failing to notice the cue, or diagnosing these cues incorrectly, is illustrated in our medical case study in Chapter 6, 'Lessons from the NHS'.

What shall I do?

There are four principal methods involved in making decisions. These are:

1   recognition-primed
2   rule-based (or procedure-based)
3   choice through comparison of options
4   creative.

RECOGNITION-PRIMED DECISION (RPD)

These decisions are based on remembering the responses to previous, similar situations. It is a fast decision-making process, sometimes called 'intuition' or 'gut feel'. The outcome is a satisfactory solution to control the situation, even if it is sub-optimal.

RULE-BASED

Following diagnosis of the problem, the employee follows the rule book on what to do. High-risk activities are governed by the rule book and operators are required to consult the manual before taking action. Decisions are thus slower, unless the rule has been learned. Rule-based decision making is also helpful for operators when justifying their decisions and actions after the event.

CHOICE THROUGH COMPARISON OF OPTIONS

Analytical decision making concerns developing a choice of actions which are compared in order to determine which is the most appropriate in the situation. If an army unit is dealing with a hostage situation, the commander may have three of four options to consider by comparison.

CREATIVE

This is the basis of innovation in an unfamiliar situation, exemplified by Captain Hayes in US Airways flight 232. We discussed NASA, reminding readers of Apollo 13 in Case Stidy 4.1, where the crew and mission controllers in Houston had to work together to solve a series of problems.

Factors in decision making

Competence in decision making is influenced by experience, technical expertise, familiarity with the situation and rehearsals and drills in emergency situations. Other factors which can impact upon the quality of decision making include:

- stress (tunnel vision, speed/accuracy trade-off)
- fatigue (selection of simple solutions over complex ones)
- distractions
- correct/incorrect situation assessment
- jumping to conclusions
- not communicating/failure to consult
- unwillingness to challenge 'the experts'
- assuming you don't have the time
- failure to review past experiences.

Training decision makers

DODAR AND FORDEC

Two simple acronyms used in providing training to decision makers are DODAR and FORDEC. DODAR is used by British Airways:

D – Diagnosis: what's the problem?
O – Options: what are the options?
D – Decision: what are we going to do?
A – Assign: who does what?
R – Review: what happened?

FORDEC is used by Lufthansa:

F – Facts
O – Options
R – Risks and benefits
D – Decision
E – Execution
C – Check.

Tactical decision games and simulators

Tactical decision games and simulators provide opportunities for practice. Formula 1 teams have simulators for every race circuit in the calendar which replicate every turn, kerb and surface change. Drivers spend hours in the simulator before races so that they are fully prepared when they arrive at the track. Police forces use simulators in streetcraft training which allow officers to make decisions on scenarios. Depending upon their decision, events unfold to their conclusion. After the exercise, officers are debriefed to capture the learning. Energy companies use simulators to practise drilling techniques for oil and gas reservoirs. Tactical decision games are used by the army and others. They comprise brief, written scenarios designed to exercise non-technical skills, especially decision making. The scenario is prepared by an expert to contain ambiguous or missing information, culminating in a dilemma to be solved in short time. Again, this provides learning for the participant and the organisation from thoughtful debriefing.

Story-telling

All organisations have 'story-tellers' – those who tell stories about this and that which happened in the past. This is a powerful way to help people make sense of events and share lessons. Klein (1998) argues the power of stories as a vicarious experience for those who were not there.

The time interval between NASA disasters in 1986 and 2003 represents perhaps half a career. Many of those who were there in 2003 were not there in 1986. Story-telling provided continuity for those in the first half of their careers.

Case study 4.2 Lessons from a Fire and Rescue Service

Deborah Clements was the Strategic Risk Manager, and part of the senior management team known as the Policy Advisory Group, at Staffordshire Fire and Rescue Service from 2003 to 2005. She has considerable occupational health and safety experience, both before and since, gained in occupational health service provision, construction and the emergency services. She was Head of HSE Assurance at CLM during the construction of the London Olympic Park from 2009–11.

During her time with the Fire Service, Deborah's role involved contribution to policy decisions on resources, service expectations, and in heat maps and planning decisions in Staffordshire.

Clements (2013) summarised the drowning of Sub-Officer Paul Metcalfe of the Greater Manchester Fire and Rescue Service, and the subsequent legal processes. These, she says, provided considerable learning for her management team.

> Paul Metcalfe (Figure 4.6) died on 5 September 1999 as he tried to rescue a teenage boy from a small lake in Holcombe Brook, Bury. Reyaz Ali, 15, had been with friends when he swung on a rope into the water and drowned.
>
> Mr Metcalfe, who was based at Ramsbottom fire station, was part of a rescue operation to try to reach him. His lifeline became snagged underwater as he tried to pull the boy to safety.
>
> Mr Metcalfe's family made an official complaint against Barry Dixon, Greater Manchester's Chief Fire Officer, alleging he was neglectful in failing to prevent the fire-fighter's death. A six-month inquiry by the Commissioner of the London Fire Service cleared Dixon of any neglect in the death of Mr Metcalfe from Ramsbottom. The conclusion was described as a 'whitewash' by the solicitors representing the Metcalfe family.
>
> The brigade's chief fire officer, Barry Dixon, had led a review of waterborne rescues shortly before Mr Metcalfe's death.

An inquest later recorded an open verdict into his death. His family subsequently failed in an attempt to press the Crown Prosecution Service to bring a charge of corporate manslaughter against the Service.

The Service was later prosecuted by HSE in 2004 over the events. The court was told that for five-and-a-half-years before the tragedy, internal memos had raised concerns about the need for water rescues to be made safer for firefighters. But after deliberating for three-and-a-half-hours, the jury found Greater Manchester Fire and Civil Defence Authority not guilty.

Since Mr Metcalfe's death, fire tenders in Manchester have been equipped with two life jackets, floating lines and inflatable hoses. All of the Services' ~2000 frontline firefighters have since received specialist water rescue training.

My management team maintained a close watch over this tragedy, reviewing the investigation reports, the legal papers, and seeking to learn lessons for our own Service. The Fire Brigade Union pressed for 'a scalp'. The whole event provided considerable learning for our management team.

Figure 4.6
Sub-Officer Paul Metcalfe

Authors' comment: strategic–predictive–dynamic

We observe the following characteristics in relation to our 3-Level Risk Management Model™ shown in this chapter.

1 Strategic: There was provision of new equipment and new training, but only after the event.
2 Predictive: There appears to have been five-and-a-half years' opportunity to address the concerns for water rescues to be made safer within the Service and to solve them in slow time.
3 Dynamic: When an incident occurs, everything becomes fast time; when the proper rescue equipment and training are not available, humans (firefighters) often follow the natural urge to rescue. They improvise, and on this occasion a rope around the waist was insufficient.

Case study 4.3 Medium-pressure gas main strike

On 18 March 2013, some roads in Enfield were being planed to a depth of 100mm as part of the night's resurfacing works. At 23.50 hours the tracked planer was about to complete the final cut of the night when it hit a medium-pressured gas pipe which had only 50mm depth of surface course covering it, as shown in Figure 4.7. All works stopped immediately, as a large amount of

Figure 4.7 Medium pressure gas main ~50mm below the road surface

---

gas was escaping from the pipe. The supervisor called the National Grid and the Fire and Rescue Service.

Gas response team arrived at 00.15, with the Fire Service attending at 01.00. Police were also in attendance as the whole area was evacuated (approximately 100 residents), with extended road closures put in place and the Enfield railway station also closed.

At 03.00, a police 'Gold' commander arrived on site to take charge of the incident. The gas leak was eventually stopped at 04.30, as the gas utility company had difficulty identifying the correct valve to turn off without affecting supply to large areas of the town.

Resurfacing work resumed at 04.30, and the road reopened to the public at 10.30. The railway station was reopened at 06.30. Other roads remained closed while the gas utility company set about lowering the main pipe.

Authors' comment: strategic–predictive–dynamic

We observe the following characteristics of failure in relation to our DRA model shown in this chapter.

1   Strategic: High-level policy to keep roads in good order led to planned maintenance of the highway. Distribution Network Owners (DNOs) maintain drawings of the location and position of their assets, including gas mains.
2   Predictive: There was no one from the principal contractor on site to supervise and monitor operations, or to check that policies and guidance to protect underground services were followed. Operational assessments relied upon the DNO drawings of services. National policies expect a gas main of this size to be at least 600mm below the road surface
3   Dynamic: The gas main was unexpectedly shallow below the road surface, potentially leaving it open to damage from heavy vehicles traversing or maintaining the road. Working methods often vary at night, when there are fewer (or no) managers around. The operators on site did not use locating devices or make trial digs to test the conditions on site (Figure 4.8).

Regulator's guidance (HSG47) describing the safe procedures and steps to be taken prior to working near to underground services is available for download from this book's companion website www.routledge.com/cw/dynamic-risk-assessment.

Figure 4.8 'Let's not bother with trial pits'

All of our examples and case studies serve to illustrate how the risk assessment loop for integration and continual improvement can be achieved in practice.

Chapter summary

Once again, we'll give you an opportunity to check your knowledge before you move on to the next chapter.

1 What are the three levels for making decisions and assessing risks?
2 Who is the only member of management to communicate directly with the crew of a manned space flight?
3 For seven consecutive years following the introduction of ORM in the fire and rescue services, what data suggested that the new approach was probably successful?
4 What are the two stages in the process of making a decision?
5 In the context of decision making, what are Gold, Silver, and Bronze?
6 What is DODAR?
7 What is FORDEC?
8 In which industry did CRM (crew resource management) emerge?
9 What is ISO 31000:2009?
10 What is the value of story-telling to an organisation (in the context of this chapter)?

How to embed The 3-Level Risk Management Model™ within an organisation

Tell me and I'll forget; show me and I may remember; involve me and I'll understand.

(Chinese proverb)

In Chapter 4 we introduced The 3-Level Risk Management Model™. We discussed the principles of PRA undertaken in slow time, and when applied in the field in fast time. We explored ways in which organisations can enable their workers to make effective, risk-based decisions in the field.

In this chapter, we explore some key areas which we believe need to be in place for any organisation seeking to adopt dynamic risk assessment (DRA) as a risk management tool. In doing so, we provide organisations that currently use DRA with an opportunity to review and revise their existing arrangements as may be necessary, as part of continual improvement.

The topics that will be covered include:

- securing senior management support
- challenging organisational culture
- core values
- safe person concept (employer and employee role)
- tips, examples and solutions.

Introduction

DRA is fast becoming an approach adopted in the public, private and third sectors, extending wider than its original roots. DRA is recognised by the HSE and the courts. The latter are developing understanding of DRA through case law, as referred to in Chapter 1 (Case study 1.2).

We have described earlier how the practice of DRA has spread (web hits, case law and wider non-emergency use) over the years, with many organisations applying this risk-based methodology. Fundamentally, the risk management framework within your organisation needs to be supportive of your service delivery, while maintaining focus on making people safe. This will aid workers at all levels within the organisation to make the right decisions.

Securing senior management support

Securing senior management commitment to occupational health and safety has become a perennial challenge for health and safety advisors and non-specialists alike. The traditional business case for securing health and safety engagement was with financial, moral and legal drivers. We see a need for a new approach in order to drive through the desired changes so as to continually meet organisational objectives and comply with the law. The method of engaging senior management requires renewing. If health and safety practitioners wish to enjoy the support their subject deserves, they need to learn to speak the language of the board room. A strong understanding of The 3-Level Risk Management Model™ will help them, by enabling senior managers to see the connection with their strategic decisions and how these enable PRA to support operational delivery and risk-based decision making in a changing environment.

Senior management teams are gradually tuning in to the value of a broader approach to risk management, and the health and safety profession must be ready to rise to the challenge of enabling this change to take place.

Securing senior management engagement in DRA will be driven by the nature of the business strategic objectives, its risk profile and the internal and external expectation for the risks to be managed.

The board and the Health and Safety Champion

The Institute of Directors (IoD) says that whilst the chief executive can give the clearest visibility of leadership, boards will find it useful to name one of their number as the Health and Safety Champion. We go further; our fifty plus years of experience, magnified by the lens of research undertaken for this book, suggest that this good practice is actually a core action for boards. For many organisations, health and safety is a corporate good governance issue and boards should integrate health and safety into the main governance structures and sub-committees of risk, remuneration and audit.

The Turnbull report on the combined code on corporate governance requires listed robust systems of internal control (Financial Reporting Council, 2005). These systems of internal control should cover risks relating to the environment, reputation and health and safety as well as financial risks. The IoD guidance (IoD, 2009) is available to download from this book's companion website (www.routledge.com/cw/dynamic-risk-assessment).

The board's Health and Safety Champion has a key role to play in any organisation, and securing their support for the adoption of The 3-Level Risk Management Model™ will greatly enhance the opportunity for DRA to be incorporated within the organisation's risk management structure.

Subject matter experts, advisors and specialists who understand the business need to be working collaboratively with key stakeholders to ensure that senior managers engage and provide leadership in embedding this structure. The concept of the 3 levels in the risk management model was illustrated in Figure 4.1: this must be aligned to support your organisation's structure(s).

The success (or otherwise) of your DRA implementation will depend on how well it has been embraced by senior management and then integrated within the organisation's health and safety management system.

If we consider the 3 levels of risk management starting at the dynamic level, and assume for an instant that there is nothing else, then this approach is doomed to fail at some point in the future.

> Sometimes, when I talk with boards, I feel like the boy who cried wolf. I do, however, remind the board that actually at the end of the story the wolf really does turn up.
>
> (Stephen Asbury)

For a powerful example of what can happen when the connectivity between the three levels of risk management is broken, we encourage you to read/reread the conclusions from the Hillsborough disaster on pages 57–61.

As we have said, any organisation adapting or adopting The 3-Level Risk Management Model™ must have an understanding of the relationships between strategic, predictive and dynamic risk assessment. If your organisation is like-minded, you will find it rather easier to complete the transition to this approach.

This transition will reinforce the health and safety culture and strengthen its legitimacy, and whilst we recognise that this new culture will prove challenging for some in your organisation, we suggest that there is no alternative.

TIP

Make sure that the collective accountability in the board room for health and safety is not diminished by the appointment of the health and safety champion.

Example: An organisation appointed a HR director as Health and Safety Champion and the other directors behaved and spoke as though it was no longer their individual and collective responsibility, as 'HR deal with this stuff'.

Solution: Ensure that policies clearly articulate and define individual and collective accountability.

Culture challenges

Ravasi and Schultz (2006: 437) define organisational culture as 'a set of shared mental assumptions that guide interpretation and action in organisations by defining appropriate behavior for various situations'.

Safety culture is a subset of organisational culture, and is defined as 'the product of individual and group values, attitudes, perceptions, competencies and patterns of behaviour that determine the commitment to and the style and proficiency of an organisation's health and safety management' (HSC, 1993, cited in HSE, 2005).

It is well documented that a poor safety culture leads to accidents at work. This same culture (positive or not) is a good indicator for determining how effectively DRA will be implemented. Organisations that adopt DRA must have a positive safety culture and genuine commitment by senior management.

TIP

Make sure that middle management 'get it'.

Example: A large international organisation reported having a less than admirable safety culture. When we talked with the senior management, we were persuaded of its genuine commitment to high standards and cultural development. When we attended the health and safety committee, the operations staff made clear their wish for better and safer working conditions.

All factors pointed to middle management being very busy; indeed too busy fighting fires on delivery dates and product quality and attending meetings, and health and safety was crowded out of their agenda.

When we visited the toilets and kitchens they were filthy.

Solution: Senior management should make weekly safety tours, asking pointed questions of middle managers and expecting coherent answers on health and safety matters.

Challenging cultures

There are many non-emergency occupations that are challenging and demanding, such as the armed forces, teaching teenagers or working on an NHS ward. In the emergency services, the role of a police officer is a difficult one when standing between the law-abiding citizen and a criminal. No matter what action they take, it can lead to criticism from either side.

Sterling (1972) found that the job performance and behaviour of a police officer is influenced by a number of factors, including public expectation, government agendas, service procedures and line management instructions. The research found that the officers perceived conflicting expectations from different audiences, and faced a range of complex behavioural expectations which tended to define their role and structure their performance in that role.

Oligny (1994) found that people who join the Police Service have strict codes of ethics, coupled with a heightened sense of duty that translates into enthusiasm, dedication and a strong emotional commitment to work. Likewise, Sterling (1972) reported that police officers have a certain attitude and behaviour in the way they operate, which is developed when officers join as cadets. This is reinforced over time, such as the goal to achieve operational objectives and to fight crime. Police officers see this as their purpose in order to make a difference. This behavioural response is typified by an officer going against the grain during an incident and not thinking of 'their own' safety, instead considering the wider objectives of operational policing (Home Office, 1996). An example of this is a case involving a male suspect wielding a knife whilst being chased by officers (wearing stab-proof vests). Detective Constable Swindells decided to leave his office without a stab-proof vest to join the chase and assist his colleagues, and pursued the knifeman on a narrow canal bank. He was fatally wounded by the suspect. The Chief Constable of West Midlands Police said: 'the first duty of police officers is to preserve the lives of the public and I think it would be fair to say that he did that' (Harrison, 2004). Line managers have a key role to play in impressing upon their teams the importance of personal safety at times when this may be compromised by other people's expectations and a sense of duty that may drive them to immediate or irrational action.

TIP

Looking at this from the point of view of your own organisation, what are the equivalent scenarios that might affect behaviours within your team? It is important that you recognise these issues and clarify your expectations.

Case study 5.1 Not looking for dead heroes

There may be a public and media expectation which is not aligned with the way that organisations operate. Example: 'Coastguard sacked for daring cliff top rescue of 13-year-old schoolgirl' (Mail, 2008). The case concerns a coastguard's actions and media interest following an incident which led to negative health and safety publicity. The coastguard officer decided to go to the aid of a schoolgirl who was trapped on a cliff. He said: 'I couldn't have lived with myself if I had waited any longer and she had fallen to her death.'

He did not wait for back-up, saying that he did not have time. He climbed down the cliff without ropes or a harness to hold onto the girl until assistance arrived. The Maritime and Coastguard Agency (MCA) argued that the officer did not adhere to operating procedures. A spokesman said: 'The MCA is not looking for dead heroes. As such, we ask our volunteers to risk assess the situations they and the injured or distressed person find themselves in, and to ensure that whatever action they take does not put anyone in further danger.'

Figure 5.1 'We don't want dead heroes'

Authors' comment

We observe the following learning from the MCA case study. It is clear that in this case the organisation had different expectations to at least one member of its staff. One of the problems in cases like this is public expectation and a person's individual views may differ from an organisational view. As we have said, there can be immense pressure on individuals to take a course of action without regard for their own safety, in cases where they believe the public would expect them to act or where they hold strong personal beliefs. In Chapter 3 we discussed the psychological explanation for this, known as 'cognitive dissonance'. Case study 5.1 provides a very good example of this, where the MCA officer placed himself in harm's way so as to aid the child.

Core values

Core values are typically articulated within an organisation's mission statement and are a sign-post for the way an organisation wants to conduct its business. Core values can be described as operating philosophies or principles that guide an organisation's internal conduct as well as its relationship with the external world.

An organisation's core values encapsulate its purpose at any time, but can develop as the overall culture develops. For example, the Fire and Rescue Service articulated its core values in 1998, in its behavioural approach to risk taking (Figure 5.2). This can be described as:

> We may risk our lives a lot, in a highly calculated manner, to protect saveable lives.
>
> We may risk our lives a little, in a highly controlled manner, to protect saveable property.
>
> We will not risk our lives at all for lives or property that are already lost.
> HM Fire Inspectorate (1998)

The latest revision to these core values says:

> In a highly calculated way, fire-fighters:
>
> Will take some risk to save saveable lives
>
> May take some risk to save saveable property
>
> Will not take any risk at all to try to save lives or properties that are already lost.
> Firenet (2013)

Figure 5.2 Fire-fighters' risk appetite

The Fire and Rescue Service's core values express the importance that it places on the level of risk it is prepared to take and in what circumstances, or its 'risk appetite', in order to improve its effectiveness. The 2008 review of the *Fire Service Manual* emphasised the Service's vision and values of leadership in a model named 'Aspire', which has at its heart four core values, which are:

- service to the community
- valuing all our employees
- valuing diversity in the Service and the community
- valuing improvement.

This represents a statement of intent, a shared understanding of risk and the vision to improve effectiveness.

What do your organisation's core values say about risk?

TIP

Find out what your organisation's core values say or imply about risk.

Core values must be well understood by the entire organisation, and need to define the boundaries in which its workers operate. Think of core values as like a stick of 'Brighton rock'. If you were to break it into pieces you would see the Brighton rock symbol repeated in every section, wherever it is broken. An organisation's core values and expected behaviours should be exactly the same and should permeate all levels and areas of the organisation. If the core values are fully embedded, this will aid in providing the workforce with a shared understanding of those expectations. The performance targets should be set against the behaviours and expected outcomes.

A manager may unwittingly demonstrate conflicting beliefs in the core values – by stating that people's health and safety is important; but, when under pressure to meet targets the same manager may overlook safety.

For example, in a team meeting a manager told his team that their safety on the road was top priority. However, less than two months later, when a large customer insisted on an urgent delivery, the manager demanded that a driver complete the delivery within the hour – regardless of the fact that the driver was over 70 miles away from the destination.

An organisation's core values need to be realistic and meaningful in the real world. An integral part of core values is the messages expressed as behaviours by managers to its people.

What do the following non-verbal messages say about an organisational culture?

- Eye protection is mandatory, but the manager goes around the factory floor without protection.
- The manager says health and safety is important but she does not attend the health and safety committee, due to a conflicting appointment.
- The manager says health and safety is important, but when you go to lunch you see her driving in her car whilst talking on her mobile phone.

Organisations need to keep it real and consider what the workforce will be judged on. Will it be meeting performance targets, keeping themselves and colleagues safe, or both?

Safe person concept

The law in the UK requires employers and others to think about, and deal with, so far as is reasonably practicable (SFAIRP), any occupational health and safety risks before people undertake their duties. The employer, however, cannot always control or make the place of work safe. There is still a duty to ensure that risks are properly assessed and managed. This risk assessment will identify safety-critical aspects so as to ensure operational safety and effective management. This can be referred to as the safe person concept (Table 5.1).

Table 5.1 Safe person concept

Employer responsibilities	Employee responsibilities
Personal protective equipment	Vigilant for his/her own safety and for safety of colleagues and others
Training	Able to recognise his/her abilities and limitations
Selection of staff	Competent to perform tasks assigned
Safe systems of work	An effective member of a team
Supervision	Self-disciplined to work within accepted systems of work
Effective instruction	Adaptable to changing circumstances
Provision of risk information	
Provision and use of equipment	

Source: Adapted from Home Office (1998)

Safe person concept (employer/employee)

The 'safe person' concept (see HM Fire Service Inspectorate, 1998), requires organisations, based on predictive risk assessment, to provide appropriate means to ensure the safety of their workers. It also requires managers to assure themselves that their workers can identify hazards, are involved in the risk assessment process and are competent to discharge their duties in accordance with the PRA. Staff need to be trained to a consistent and appropriate level (in line with the hazard profile) where they are competent to assess risks in the field, and to be empowered to implement appropriate controls. A good understanding of the principles of DRA is required (see Chapter 7). Training should provide a link to the principles and use of The 3-Level Risk Management Model™. Managers and staff who have not received this training will not have sufficient knowledge or skill to contribute to any risk assessments, give and receive feedback or take responsibility for others' safety in the field.

Line managers have a key role to impress upon their workers the importance of personal safety whilst maintaining efficient delivery of services. A non-exclusive list of what your managers should be doing is shown in Table 5.1

The best models for DRA are those that are simple and adapted to suit your organisation. Our recommendation is to follow The 3-Level Risk Management Model™ (Figure 4.1).

Case study 5.2 Familiarity breeds contempt

Elaine is a care worker for an NHS Trust. She has been handed the care of a new client, named Jack, from another care worker. In the handover process, Elaine was told that the client (who has mental health issues) is 'as good as gold' and has never caused anyone problems in the past. The client has just moved into his new home and he is doing well on his medication. Elaine has previously visited Jack on two occasions, first with the care worker from whom she took over and subsequently on her own. Jack appears to be adjusting well in his new home. Then Elaine gets an email message from a colleague saying that Jack has injured himself. Elaine feels that it is a good time to visit him to reassess his requirements.

What happens next?

But this time things were different!
Jack had not taken his medication regularly for the last two months and had missed two appointments with his specialist to discuss his treatment. He had been secretly self-harming. Once Elaine was in his home, he lashed out violently when she asked him about the bruises on his body (Figure 5.3). As a result of the assault, Elaine has been on sick leave for over a month, having suffered flashbacks as a consequence of her traumatic experience.

Figure 5.3 Assault on care worker

Authors' comment

In this case, there are lots of learning points for both Elaine and the trust. Elaine will undoubtedly have been relieved to have got out of the client's home alive and she may be thinking – 'why me?' She may also be thinking whether her training, information and intelligence equipped her to make the right decisions. Elaine's organisation will want to learn lessons from the investigation very quickly, so as to avoid this situation happening to someone else. As part of the investigation, the trust will also want to look at a number of issues in terms of its internal controls. This may include the PRA for this client (which is generally retained with the case notes), the standard operating procedures informed by the PRA, the information provided to Elaine on the client's risk profile and the effectiveness of information exchange between the other agencies involved with Jack.

It is important to use PRAs to inform practice in the field. This case study illustrates that there is a tendency for people to under-estimate familiar risks and over-estimate unfamiliar risks. You need to expect the unexpected when undertaking routine tasks, and be alert to the dangers of complacency. Acting without gathering all the relevant information can be dangerous. Be aware of routine and consistent patterns of work and how events can arise, so as to avoid the danger of not looking for new information. You need to use your cognitive abilities continually to assess risk. Workers need to know what to do if things do not progress according to plan, and of course this will be documented in the PRA (Jacobs, 2010a, 2010b). An example of a technique for dealing with situations such as this are described in Case study 7.2.

A risk of being supplied with client information is that you may have a tendency to be biased in your judgement. For example, in anchoring bias would mean that you rely on the initial part of the information, that Jack is harmless, when making future decisions about Jack (review if necessary in Chapter 1). We are more likely to adjust our decisions around the view that Jack is harmless rather than to consider a worst-case scenario, even when the evidence points to the contrary.

TIP

Look at your own organisation to identify how it manages its occupational health and safety risks. Is it effective?

Summary of key management points:

1   How are you going to ensure, so far as is reasonably practicable, that your staff present the appropriate behavioural responses in dynamic situations?
2   Do you have suitable and sufficient risk assessments in place for predictable risks?
3   How does your organisation prepare its people prior to and during the event/activity, and what is your contingency if it goes wrong?
4   Are your health and safety systems resilient, and proportionate to the predicted risk?
5   As an accountable manager, what expectations do you place upon your people to deal with situations that are dynamic?

TIP

Seek senior management support before rolling out a 3-Level Risk Management Model™ approach in your organisation.

Example: A line manager decides to adopt DRA within his production department without the approval or knowledge of senior management. The manager is on holiday and when he returns he is shocked to find out that a production engineer (PE) has been given a stiff talking to for not following procedure. The PE told the manager that he had used DRA to inform his decision and stopped the production line for maintenance work without notifying senior management. He believed he had made the right decision and was able to adequately explain his decision process.

Solution: The manager needs to gain Board approval for changes to organisational risk management systems, and to seek advice from specialist advisors. This will ensure that considered changes to risk strategies are fully endorsed and embedded so as to include risk assessment, escalation and decisions that impact upon production line output.

TIP

Integrate DRA methodology within the current risk management system and structures.

Example: A company has implemented DRA as a stand-alone procedure, and a member of staff suffered a limb amputation. Following an investigation, it transpires that a similar incident occurred two years earlier. This latest incident occurred six months after the introduction of DRA; however, the company did not have set procedures and the information on the previous incident was not escalated to line managers or workers, as it was seen as a separate matter. As a result, lessons were not learned and, whilst the information was recorded under the duty to notify the HSE, it was not considered by the organisation as a wider risk management issue.

Solution: The introduction of DRA methodology must be integrated within the health and safety management system and aligned to business and risk management structures so as to enable effective risk escalation to operational and/or strategic level to determine resources or policy review.

TIP

Ensure that your workers clearly understand the organisation's expected behaviours and outcomes when operating in the field.

Example: During the night shift, a security guard at a hospital is called by a nurse to assist with a patient who is known to the staff to be an alcoholic and has made lewd comments and has once in the past touched a nurse inappropriately. The nurse recognises that the patient is drunk and may also be under the influence of drugs. She is nervous about dealing with him. The security guard attends, but after about ten to fifteen minutes the agitated patient appears to calm down and the security guard goes out for approximately five minutes. Upon his return, the nurse is very upset that he left the ward and complains. The security guard argues that the nurse was over-reacting, as nothing had happened whilst he was away, and that prior to leaving he had done a DRA, arguing that the patient was asleep and he felt it was safe to leave for a few minutes.

Solution: The guard needs to follow the hospital's standard operating procedures, which would highlight the need for the patient to remain supervised and that the security guard should not have left his post. The claim that he undertook a DRA is not appropriate in this situation.

TIP

Make expected behaviours explicitly clear to workers and reinforce the message that they will be held to account for how they make decisions based on risk.

Example: A police response vehicle is responding to an incident involving an armed robbery at a high street bank, where at least one casualty has been reported. The response vehicle gives chase to the getaway car during peak shopping hours and in an area where there are a lot of pedestrians. The stolen vehicle suspected to be involved in the robbery jumps a red light, striking an oncoming cyclist. The police vehicle in pursuit continues at high speed, and during the chase narrowly misses an oncoming lorry, striking a parked car but managing to keep the suspect in sight.

Solution: Organisations need to be aware of occasions in which staff may exceed risk expectations and have in place a contingency plan. In this case control room staff have a key role in DRA and should instruct the driver to abort the chase if there is imminent danger to the public. Based on an assessment of risk, can an alternative course of action be adopted?

TIP

Ensure that all workers at all levels understand your core values and how each can use these to guide decision making.

Example: A bus driver is on duty during the early hours of the morning and she drives past a group of young men who are running after a young man and notices that some have knives. The driver does a DRA and believes that the boy who is being chased is far enough away that she can quickly stop to let the boy jump onto the bus. She screams at the boy to jump on, and then misses the next three stops to ensure that all her passengers are safe, whilst radioing for police back-up.

Solution: The bus driver has used DRA to ensure the safety of a member of the public who was on the pavement, even though he was not a passenger. The bus company should recognise that the actions of this driver may have saved the life of the young man and that the driver considered passenger safety in making her DRA. The learning from this incident should now be fed back to inform other drivers.

Chapter summary

1   Identify two safe person concepts from an employer's perspective.
2   Identify two safe person concepts from an employee's perspective.
3   Why is aligning your risk management processes to your organisation's core values important?
4   Why is it important for senior management to support the principles of DRA?
5   Why is the understanding of organisational culture important to understanding people's behaviour?
6   What are core values, and why do organisations use them?
7   Why is it useful for an organisation to be aware of anchoring bias?
8   Why is it useful to link performance targets with core values and operational objectives?

Chapter 6

Using dynamic risk assessment to improve service delivery

When a patient is acutely unwell and presents to hospital, or deteriorates and becomes acutely unwell whilst in hospital, time is of the essence and a fast and efficient clinical response is required to optimise clinical outcomes. Current evidence suggests that the triad of i) early detection, ii) timeliness of response, and iii) competency of the clinical response, is critical to defining clinical outcomes.

RCP (2012: 18)

12.45 Mr. Kipper – 37 Shorrolds Road O/S
(Suzy Lamplugh's desk diary for
28 July 1986, in Murdermap, 2013)

Introduction

In this chapter we will explain how to use dynamic risk assessment (DRA) to improve service delivery. This is not a chapter (nor a book) about fiddling with the health and safety paperwork. When we say 'improve service delivery', we mean:

- you survive your surgery
- your aircraft lands safely
- no one dies on the school trip
- the gas main outside your house does not leak or explode
- the fire-fighter survives the water rescue
- the care worker is not assaulted, losing over a month off work
- and so on . . .

We will reinforce the benefits of utilising a sensible and proportionate risk-based approach to achieve service delivery and effectiveness. The chapter will address the common issues of:

- risk-taking behaviour
- risk perception
- red mist (or 'rushing in').

Improving service delivery

Throughout this book, we have provided examples and case studies of organisations which have adopted (or failed to adopt) risk management models aligned to ours (and others') with the intention of meeting their objectives.

This chapter is specifically about improving service delivery, so let's start by looking at some specific examples and lessons to be learned from the English National Health Service (NHS) in a case study.

Case study 6.1 Lessons from an NHS Trust

Dr Craig Stenhouse has worked as a medical doctor since qualifying from the University of Leicester in 1986. He has worked in several NHS hospitals across the UK, and became an intensive care consultant in 1998. He is presently the Medical Director at the Burton Hospitals NHS Foundation Trust in Staffordshire, UK.

Stenhouse (2013) described how a well, fit young mother-to-be can become sick very quickly, and how a care team must be assembled within 4–5 minutes to save her life.

> A team looks after a mother entering maternity care. Bleeding during childbirth can be normal, or arise from performing a C-section for delivery. If the placenta is in the wrong place, a mother can lose blood very quickly, resulting in a massive obstetric haemorrhage. In our hospital last year, there were around 3600 births. We had three (3) such haemorrhage events during births – effectively, a 1:1200 risk. Whilst there may be signs which give our medical teams clues that a pending birth may be higher risk, such as a low-lying placenta, we don't know when the next case will be. We have very clear procedures to follow, which may result in hysterectomy, and we have to move very quickly. To make sure everyone knows what to do, we carry out drills and exercises so that we're ready for the next occurrence.
>
> Over the years, we have developed a series of 'early warning systems' that have helped our nurses and doctors to communicate on the signs and symptoms of illness, and quantitatively assess whether patients are

---

improving or deteriorating. 'MEOWS' is the Modified Early Obstetric Warning System used in our hospital.

Observing great respect for patient confidentiality, Stenhouse explained the evolution of the National Early Warning Scoring system, 'NEWS', from a system developed in the Burton NHS Trust. NEWS was recommended to UK hospitals by the Royal College of Physicians (RCP) as an aid to clinical assessment (and not as a substitute for competent clinical judgement) in 2012.

> In February 1998, we admitted an elderly but fit male for a second bowel operation to treat a cancerous condition. After the operation, at 6pm, he was out of theatre, OK, and having had an epidural for pain relief. By 9pm, he had become unwell. His blood pressure was down, and his heart rate elevated. A house officer (a junior doctor) was called who in turn took advice from a more senior doctor at that time working in theatre at 11pm. By 3am, he had a normal heart trace and the doctor diagnosed a heart attack based on the earlier signs. This decision was wrong. The patient bled internally all night, and died from blood loss.

> A 3–4 year investigation into those events led to an early warning system (EWS) being developed in the Burton hospital, based on a model used in the Norwich and Norfolk Trusts. This modified EWS (MEWS) allowed us to detect the need for earlier interventions, which led to patients being less sick.

> Eighty NHS trusts came to see us, as MEWS was so effective. In 2012, I wrote the national guidelines for the new National Early Warning Scoring system – called 'NEWS'.

NEWS is based on a simple scoring system in which a score is allocated to physiological measurements already undertaken when patients over the age of sixteen present to, or are being monitored in, hospital. Six simple physiological parameters form the basis of the scoring system:

i   respiratory rate
ii oxygen saturations
iii temperature
iv systolic blood pressure
v   pulse rate
vi level of consciousness.

A score is allocated to each as they are measured, the magnitude of the score reflecting how extremely the parameter varies from the norm. The score is then aggregated to provide a discriminating risk of acute mortality and a trigger level for a clinical alert.

NEWS standardises the assessment of acute-illness severity when patients present acutely to hospital, and also in the pre-hospital assessment by primary care and the ambulance services. NEWS can also be used as a surveillance system for all patients in hospitals, tracking their clinical condition, alerting the clinical team to any clinical deterioration and triggering a timely clinical response.

To facilitate standardisation and a national unified approach, a colour-coded clinical chart has been developed to record routine clinical data and track a patient's clinical condition. This tracking system will alert the clinical team to any untoward clinical deterioration, and also clinical recovery. This in turn should determine the urgency and scale of the clinical response.

Stenhouse says that this system has now become a part of hospital vocabulary 'he has a NEWS score of X'. He says that it has three main benefits:

- it empowers nurses
- it is objective, not subjective
- it standardises charts and patient information.

The full report of the NEWS working party can be obtained from this book's companion website (www.routledge.com/cw/dynamic-risk-assessment). The RCP (2012) states that high-quality versions of the NEWS charts and their explanatory text are available to download, photocopy or print direct from its website at www.rcplondon.ac.uk/national-early-warning-score. The RCP asks that lower-quality versions of the charts shown in the report itself should not be used. The charts must be reproduced in colour and should not be modified or amended.

Common issues in service delivery

We have identified three common issues which impede successful service delivery:

- risk-taking behaviour
- red mist (or 'rushing in')
- risk perception.

Risk-taking behaviour

In general, people think of risk-taking behavior as participation in activities that only courageous or 'thrill-seeking' people would attempt – diving with sharks, parachuting or bungee jumping. In reality though, risk-taking behaviors will also include more mundane acts, like having unprotected sex, gambling or taking drugs.

Atkinson (1957) defines risk-taking behaviour as 'the motive to achieve and the motive to avoid failure'. He says risk taking influences behaviour, and that the individual taking the risks assumes strength of motivation as being a multiplicative function of motive, expectancy and incentive. This accounts for two theoretical implications:

- that performance level should be greatest when there is greatest uncertainty about outcome
- that people with strong motives to achieve will prefer extremely difficult and risky tasks, whereas those with strong motives to avoid failure will prefer easy tasks and avoid extremely difficult and risky tasks.

This matters, because 'risk' is not especially particular about whom it harms. Risk-taking behaviour includes rushing in prior to thinking.

Red mist (or 'rushing in')

> Police officers involved in pursuits are one thousand times more likely to be involved in an accident than during a routine patrol.
> (Alpert & Dunham, 1990)

'Red mist' is a term commonly used in police services to describe what might happen when an officer is determined to catch a criminal. It can be defined as a feeling of extreme competitiveness or anger that temporarily clouds one's judgement (Stephen Asbury) (Figure 6.1). ACPO (Association of Chief Police Officers) defines it as:

> a complex emotional situation affecting the state of mind of drivers who can become so focussed upon an objective or outside influence [that] their ability to accurately assess driving risk is severely reduced.
> (ACPO, 2009)

Officers in Staffordshire police (Telegraph, 2013) were told in their training that 'red mist is dynamic risk assessment's worst enemy'. They are taught that 'high risk areas' include chasing suspects, domestic incidents, road accidents, searching people or any situation involving fire or water.

Red mist, or rushing in, may prevent the successful delivery of the intended service, for example when workers are hurt.

Figure 6.1 Red mist

ACPO guidance on the management of police pursuits can be downloaded from this book's companion website (www.routledge.com/cw/dynamic-risk-assessment).

Risk perception

Risk perception is a largely subjective judgement made by people about the characteristics and the severity of a risk. The phrase is most commonly used in reference to natural hazards and threats to the environment or health, such as nuclear power.

There are several theories, which are beyond the scope of this book to explain fully, why different people make different estimates of the dangerousness of risks. In summary, three families of theory have been developed, partly discussed in Chapter 1:

1   heuristics and cognitive psychology approaches – experience-based techniques for problem solving, such as 'rules of thumb' and 'an educated guess'
2   cultural theories of anthropology and sociology – the study of humans past and present
3   social amplification of risk framework (SARF) – the process by which risks are amplified, receiving public attention, or attenuated, receiving less public attention.

But this can go wrong. Things that look simple can easily lead to errors of judgement. Let's try you, the reader, out . . .

A bat and ball cost £1.10 in total. The bat costs £1 more than the ball. So how much does the ball cost?

We believe you have just answered '10 pence'. It just looks and feels right, and yet it is wrong. Many people concede to immediate impulse. Now, go back and give it some careful thought. People are not accustomed to thinking hard, and so gut feeling takes over. The bat actually costs £1.05 and the ball just 5p – a £1 spread.

Chauncey Starr (1969) found that people will accept risks 1,000 greater in magnitude if they are voluntary (e.g. driving a car) than if they are involuntary (e.g. a nuclear disaster).

SARF suggests that people pay more attention to a train crash killing 10 people than to a typical day on the UK's roads, where 10 people will die.

Gardner (2009: 6) says that when we worry about a risk we pay more attention to it. The reverse is probably true too. Right? A meeting with a would-be house buyer in a London street during the day time? Should be a piece of cake . . .

Case study 6.2 Suzy Lamplugh

Susannah (widely known as and commonly called 'Suzy') Lamplugh was a London-based estate agent who was reported to the police as missing on 28 July 1986. Ms Lamplugh (pictured in Figure 6.2) had arranged an appointment to show a house in Fulham to someone she had referred to as 'Mr. Kipper' in her desk diary. The diary recorded the details of the appointment as '12.45 Mr. Kipper – 37 Shorrolds Road O/S' – the 'O/S' abbreviation implying that the meeting was due to be outside the property.

Figure 6.2 Suzy Lamplugh

Witnesses at the time told police that they had seen a woman resembling Ms Lamplugh arguing with a man in the street, and then getting into a car (Figure 6.3). Her own Ford Fiesta was found later that day less than two miles from where she was last seen. The ignition key was missing and Ms Lamplugh's purse was found in a door pocket.

Police investigated many lines of enquiry, including Dutch males named 'Kuiper', due to the possibility of mishearing the pronunciation, but nobody of this name was found to be connected to Ms Lamplugh. Extensive police searches in Worcestershire and the West Midlands proved unsuccessful and she has never been found.

She was officially declared deceased in 1994, presumed murdered. Re-opened investigations in 1998 and 2000 failed to uncover any trace of her.

The Suzy Lamplugh Trust was founded in 1986. Its aims are to help people avoid becoming victims of aggression, and to offer counselling and support to relatives and friends of missing people.

There was a similar case in 2006 involving a female estate agent who met a client called 'Mr Herring'. She was attacked with a knife, pushed to the ground, but managed to free herself. The assailant ran away. Police have said there is no connection between this case and the disappearance of Ms Lamplugh.

Figure 6.3 12.45 Mr. Kipper – 37 Shorrolds Road O/S

Authors' comments

We observe the following characteristics of failure in relation to our 3-Level Risk Management Model™ shown in Chapter 4.

1 Strategic: There is a fundamental and inherent perception (in this case and generally) that lone working is safe unless there is a specific reason for it not to be so. Examples in our other case studies provide instances of some of these specific reasons. In this case (and commonly elsewhere) there seemed to be no high-level policy at the estate agency relating to safe systems of work.
2 Predictive: Maintaining an accurate diary of movements and meetings is good practice, and recommended generally to lone workers.
3 Dynamic: It is unclear what happened to Ms Lamplugh.

Case study 6.3 Piper Alpha

The *Piper Alpha* oil and gas production platform, built in 1976, was located 110 miles north-west of Aberdeen. It was very productive; at its peak in June 1988, it produced 30,000 tonnes of crude oil per day, 10 per cent of the UK's North Sea oil production (BBC, 1997).

At about 22.00 hours on the evening of 6 July 1988 there was an explosion on the production deck of the platform, caused by the ignition of escaping gas from a temporary flange. The resulting fire spread rapidly, and there were further explosions.

At 22.20, there was a major explosion caused by the rupture of the high-pressure gas pipeline from the nearby *Tartan* platform, operated by Texaco.

A high-pressure gas fire raged, with additional explosions. In less than three hours, the structure of the platform had failed and fallen into the sea.

Of the 227 crew on board, only 61 survived – there were 166 fatalities that night, and a further death when a worker succumbed to his injuries days later in hospital. The great majority of the survivors escaped by jumping into the sea, some from as high as 60 metres (175 feet) (Flin, 1996).

The common image of the disaster was of a sudden, catastrophic explosion. The reality is different – during one critical hour, there were moments when it could have been averted. Dr Allan Sefton, Operations Director for the Offshore Safety Division of the HSE said:

> The explosion on Piper Alpha that led to the disaster was not devastating. We shall never know, but it probably killed only a small number of men. As the resulting fire spread, most of the Piper Alpha workforce made their

way to the accommodation where they expected somebody to be in charge and would lead them to safety (Figure 6.4). Apparently, they were disappointed. It seems the whole system of command had broken down.

(Sefton, 1992)

Lord Cullen chaired the subsequent public inquiry. Making over 100 recommendations, the inquiry report (Cullen, 1990) was critical of the manager on *Piper Alpha* on the night of the disaster, as well as the managers at the nearby *Tartan* and *Claymore* platforms, which were connected to *Piper Alpha* by hydrocarbon pipelines. It is suggested that, had production by these platforms been shut down earlier, the situation on *Piper Alpha* might have deteriorated more slowly. Some of the report is quite troubling:

> It is unfortunately clear that the OIM [offshore installation manager] took no initiative in an attempt to save life . . . in my view, the death toll of those who died in the accommodation was substantially greater than it would have been if such an initiative had been taken.

(Cullen, 1990: para 8.35)

A large part (~10 per cent) of the UK's oil and gas producing capacity was wiped out within an instant.

Figure 6.4 Piper Alpha disaster

Authors' comments

We observe the following characteristics of failure in relation to our 3-Level Risk Management Model™ shown in Chapter 4.

1 Strategic: The platform was made up of four modules, separated by fireproof walls, arranged so that dangerous works such as drilling were as far from the accommodation blocks as possible. The adaptation to produce gas (as well as oil) led to the location of gas compressors close to sensitive areas such as the control room. Blast-proof walls were not incorporated.

2 Predictive: A report with recommendations commissioned by management describing the required protection for the high-pressure gas lines was not implemented. The permit-to-work system had degraded, and was not working properly. Safety rules require crew not to jump into the sea – platforms are very high (30–60 metres), the North Sea is mountainous and very cold.

3 Dynamic: Decisions to shut down production at two nearby platforms were delayed, as managers were not empowered to make them, and communication was poor. Individual crew decisions (no doubt in fast time) to ignore their training and jump into the sea saved 61 lives.

Chapter summary

Once again, we'll give you an opportunity to check your knowledge before you move on to the next chapter.

1 Give three examples of risk-taking behaviour.
2 What is 'red mist'?
3 Give three examples of behaviours associated with 'red mist'.
4 What are 'heuristics'?
5 What did Chauncey Starr discover about the acceptance of voluntarily entered-into risks (such as smoking) and risks that had to be suffered involuntarily (such as nuclear power stations)?
6 What is SARF?
7 A bat and ball cost £1.10 in total. The bat costs £1 more than the ball. So how much does the ball cost?

Chapter 7

Training informed by strategic, predictive and dynamic risk assessment

Give a man a fish and you feed him for a day. Teach a man to fish and you feed him for a lifetime.

(Chinese proverb)

Introduction

In Chapter 6, we looked at the benefits of using DRA to improve service delivery. We explained the implications of risk-taking behaviour, red mist and how it can adversely affect performance. We also reviewed risk perception, which is a vital area in understanding how workers perceive risk and how they may interpret it in the field.

In this chapter, we explain ways in which the 3-Level Risk Management Model™ can be integrated within training. We have discussed in earlier chapters the need to adopt a risk-based approach to organisational decision making.

The topics for this chapter include:

- health and safety training
- drivers for effective training
- training biases
- case studies
- Maybo case study (1).

Health and safety training

Many organisations are looking at innovative ways to train and upskill their workforce to improve service delivery. There are a plethora of training providers and a variety of training approaches in the marketplace. Some of this learning is approved or certified and some not. Some of the trainers and instructors are professionally qualified in their discipline and some not.

Organisations spend a considerable amount of their training budget on health and safety training. In times of austerity, it is easy to see why health and safety training may be a candidate for early cuts. Our view is that a well-implemented approach to risk management will be effective in the general support of the organisation's objectives. When budgets are under greater scrutiny, providing better awareness to staff of how their actions (or inactions) can help the organisation to be successful is essential.

Organisations that have people who can effectively articulate the benefits of aligning training to the organisation's objectives, beyond legal compliance, are likely to gain greater senior management support. If organisations can truly align the training with their core values and objectives, and this is understood by senior management, you have a powerful case for the training budget.

Organisations that adopt dynamic risk assessment (DRA) within their risk management system need to look carefully at how they deliver training to those who apply the DRA concept in the field and their line managers, who have responsibility for its oversight and utility. If this is important to your organisation, then this is what your training should focus upon.

The worker needs to fully understand every PRA that applies to their role, because it tells them what they must do to control the risks. Many organisations ask their workers to sign or otherwise endorse the PRA. From the worker's perspective, this signature should not be provided lightly.

Training providers

Organisations should consider who delivers their health and safety training. As well as being a chartered member of the relevant body (see Appendix), the tutor should understand the practicalities of how it is to be applied in the field and how workers can make risk-based decisions to keep them safe. We introduced the Occupational Safety and Health Consultants Register (OSHCR) in Chapter 3, and the professional bodies that support this register assure the competence of their members and oversee members' continual professional development. Further advice on general health and safety training can be obtained from this book's companion website (www.routledge.com/cw/dynamic-risk-assessment).

Critically, organisations must satisfy themselves that their chosen training provider (both in-house or external) has practical experience, knowledge and competence in both the delivery of training and experience of applying SRA/PRA/DRA training in the field, as required by SKATE (see Chapter 3).

TIP

Resist a 'sheep dip' approach to training – one course! A one-course approach does not fit all. Organisations will need to agree on an appropriate frequency of training and any top-up or refresher training.

Example: The drowning of schoolboy Max Palmer at Glenridding Beck (Case study 2.1), emphasises the importance of personnel being skilled to make the right decision in rapidly changing environments. The HSE recommended that activity leaders should be competent in DRA. The sequence of training in SRA/PRA/DRA targeted at specific levels in the organisation clarifies what is expected of people in terms of their responsibilities and how they are empowered to take risk-based decisions in the context of their role and organisational expectations.

Solution: When considering training for PRA/DRA consider your workers in terms of how they learn and how they handle situations in the field.

Drivers for effective training

We have discussed legal responsibilities placed upon employers, and these include the provision of training.

The training and learning programme, whilst having a core structure focused on learning and developing skills, should remain up to date in line with business requirements whilst focusing on outcomes, behaviours, values and competencies.

TIP

The training should also pay attention to discussing the difference between expert and novice decision makers and why it is important within the organisation.

Where possible avoid running separate risk assessment training and DRA training; the two are inextricably linked and should be treated as such.

Managers and staff in the field who have not received training in PRA and DRA methodology will not have sufficient knowledge or skills to contribute effectively to debriefing sessions.

There are some key points that senior management may want to understand and consider, in advance of supporting a SRA/PRA/DRA training and learning programme. These include:

- the content of the course, e.g. will it cover how decisions are made in the context of occupational health and safety; human factors such as our biases and how to recognise them, as well as reconciling these against organisational values and service delivery objectives?
- what underpinning documents are going to support the learning and training; for example PRA, standard operating procedures and associated policies
- how workers can get better at what they do; the need to ensure that there are opportunities for workers to improve performance; debriefing and feedback provided by line management, and how this will facilitate worker learning
- consideration of new workers, such as inductees or existing staff who have transferred from another work area; how they will react to hazards in the field; and their ability to continually assess risks to make informed judgements about safe systems of work
- what helps managers and workers to make the right decision.

Time for change?

So, when was the last time you reviewed your health and safety training programme? We know that it is likely to consume a large chunk of your training budget.

Reinforcement of core values, discussed in Chapter 5, is not something that can be left to classroom-based training. The training in SRA/PRA/DRA should incorporate expected behaviours and be seen as an integral part of the organisation's values. If the organisation gives mixed messages about its commitment to risk management, this will become evident in how learning is delivered and received. If the learning experience is not positive, this may unintentionally undermine well-intended training. As professional health and safety trainers we sometimes come across a scenario that plays out a little like this:

Participant: This training is great, but my boss is not going to give me the time to actually do this.

Tutor: Thanks for your kind words, but do you realise that it was your boss who enrolled you onto this class?

Solution:

Senior management should ensure that all staff understand that the content and implementation of training is approved by the organisation and that it represents its core values and expected behaviour.

The training should inform behaviours when in the field. In Case study 5.1, the coastguard officer who attempted to save a child from a cliff fall was criticised by his organisation. Such situations of dealing with societal and organisational expectations must be rehearsed in training as case studies and aligned to policy expectation and standard operating procedures. When considering the context for different scenarios, whoever is delivering the training needs to be knowledgeable enough to explain low-probability, high-consequence events (which we call 'black swans').

Choosing the right scenarios

It is always useful to include 'what if' scenarios in the training toolbox. These provide a basis for delegates to explain how they would deal with a given situation, why they would deal with it in that way and how they have arrived at their decision, and what factors informed their decision making. It is also useful to include case studies that are realistic and involve situations outside of PRA and standard operating procedures.

In your chosen scenarios consider bringing examples from PRA/DRA events that have occurred. Try to introduce situations which require workers to make quick decisions and to highlight any cognitive errors and biases. This will help to illuminate that we are all subject to biases and that the more that we are aware of this, the more likely we will consider it in our decision making. For example, if you were to ask people to decide quickly how to put out a fire, and showed them many pictures of fires that correctly required water as the extinguishing agent, and also included a few pictures where it would be incorrect to use water (e.g. a chip-pan fire), they might make the mistake of selecting an incorrect extinguishing agent. This incorrect decision might be reinforced by mental images of films they have recently watched showing fire-fighters using a fire-hose to extinguish a fire. This is an example of availability bias, where people use recent images and experience to make a quick decision. Of course, this example would not work with fire-fighters, as they are highly trained in fire-fighting and would be primed with an automatic response to fire cues.

There is genuine value in health and safety cross-training with multi-agency workers where there are cross-cutting and shared activities and goals. In the same way, organisations would share PRAs as part of planning in multi-agency working. Equally, when planning such training where it requires your organisation to collaborate with another, consideration should be given to understanding different organisational goals, decision-making processes and risk priorities (which may be necessary, as part of the planning documentation exchange).

President Harry S. Truman said:

> It is amazing what you can accomplish if you do not care who gets the credit.
>
> (Truman, 33rd President of the United States of America)

The health and safety training requirement should be informed by the application of PRA for the type of environment to be encountered and be fit for purpose against the PRA requirement. Training should also be informed by feedback from debriefs, which will be discussed in Chapter 8.

Training doesn't come with guarantees and you should not aim to cover all eventualities in scenarios. You need to draw the line somewhere. Instead, a general but focused cadre of relevant scenarios should be used. This will allow for a broad range of predictable situations and types of hazards that workers are likely to encounter during the course of their work. This will enable the worker and managers within the risk-based decision-making chain to apply risk control approaches and behaviours, combined with their experience in line with risk assessments and established safe systems of work. This approach of using relevant field-based scenarios will better prepare workers and line managers, who will feel better prepared in the field.

We discussed the safe person concept in Chapter 5; this is pertinent to the employer's responsibility to provide training, as outlined in Table 5.1, to enable workers to undertake their tasks safely and efficiently. The 3-Levels of Risk Management Model™ (see Figure 4.1) require health and safety training to be provided at all levels. Workers have responsibilities at both the PRA and DRA levels and the training they receive must equip them to perform tasks competently, whether operating within teams or as a lone worker. They need to be able to know how to deal with the hazards and risks which they encounter within their normal day-to-day duties, using measures which the line manager has put in place to control the risk, based on the PRA. Additionally, the PRA should address how to deal with the emergency situations, highlighting the need to make an on-going assessment of risk in dynamic situations. The training should also equip workers with skills to adapt to a changing risk environment and to feed back (see Chapter 8) following events to verify that systems are working effectively or to review risk-based systems when necessary.

Training biases

A key element of training should include knowledge of the types of biases that individuals and groups are prone to (refer back to Chapter 1 for an outline of some types of biases). Many researchers (e.g. Baron, Badgio & Gaskins, 1986; Klein, 1998) argue that education is vital in trying to reduce errors such as myside bias, where a person finds it difficult to change a strong belief/view, even when the evidence suggests that their belief may be incorrect.

A key element that is often overlooked is the case of near misses. In many cases a serious incident or major disaster may have been preventable, but for the failure of organisations to record, review or act upon near misses. Space exploration is by its very nature a dangerous activity; technology changes, unknown hazards and human nature make this endeavour dynamic in every sense of the word. In the case of the *Columbia* space shuttle tragedy on 1 February 2003, the accident was caused by a piece of insulation foam breaking away from one of the shuttle's external tanks and hitting one wing, thus breaching the shuttle's Thermal Protection System. Sadly, upon re-entry to Earth seven crew members died after hot air penetrated the left wing, leading to structural failure and break-up of the shuttle. However, this was not the first time that pieces of insulation had separated from the external tanks. The Columbia Accident Investigation Board (2003) reported that insulation foam breaking away was a commonly reported near miss, and over time 'NASA engineers and managers increasingly regarded the foam-shedding as inevitable, and as either unlikely to jeopardize safety or simply an acceptable risk' (Columbia Accident Investigation Board, 2003: Chapter 6, p. 2). In effect, this known safety risk appeared to have been downgraded to a maintenance issue and did not receive the necessary attention it deserved.

In an attempt to separate management decisions on flight scheduling and cost from technical decisions on safety, one of the many recommendations of the Columbia Accident Investigation Board (2003: Chapter 11 (1), p. 3) was to:

> Establish an independent Technical Engineering Authority that is responsible for technical requirements and all waivers to them, and will build a disciplined, systematic approach to identifying, analyzing, and controlling hazards throughout the life cycle of the Shuttle System.

This recommendation also touches on an additional issue that can affect decision making. This is the pressure to conform to the group's beliefs, behaviours or attitudes (see groupthink bias in Table 1.2). The Columbia Accident Investigation Board (2003) found evidence that the culture within the organisation had changed; it 'now it seemed to some that budget and schedule were of paramount concern'. Workers argued that this change in organisational culture did have an effect upon their decisions:

> . . . and I have to think that subconsciously that even though you don't want it to affect decision-making, it probably does.
> (Columbia Accident Investigation Board, 2003)

The suggestion of separating management decisions on scheduling and cost from safety concerns is one way to reduce the likelihood of cost and deadlines being placed disproportionately ahead of safety. It is also an example of a strategic decision that can be implemented to ensure that risk assessment is considered at all levels. Without an integrated strategic-level approach to risk (e.g. positive organisational safety culture), latent failures may occur at predictive and dynamic levels.

During training, managers need to also consider the problem of over-confidence. Individuals have a tendency to subjectively measure their decision-making skills as higher than they objectively are. Over-confidence can lead to poor decision making because the individual may perceive the training as too easy and/or over-estimate their judgement skills. Over-confidence may result in individuals not considering all the options, or reviewing risk systematically because they are too confident of the outcome. If possible, training should try to include surprise elements to test people (throw in a curve ball) and make them aware that over-confidence can lead to nasty surprises.

Case study 7.1 Noise nuisance

Let us take an activity involving a home visit, which may have dynamic elements (unforeseeable risks) within it. The role is that of an Environmental Health Officer (EHO) for a local council. A review of the specific tasks for this particular activity will allow appropriate tactics to be applied. Once we know the tasks, we can predict the likely hazards the EHO will face. Our next set of questions might be, whether we have documented procedures and is our EHO trained? This will provide a level of assurance that the EHO can identify hazards in the field, is competent to execute their duties effectively and has been involved in the PRA process.

So, when they are out in the field, performing their duties, what is going through their mind?

An EHO goes to a house following a number of noise complaints. Other colleagues have been to the address before and, based on the evidence obtained, it is intended to serve a noise abatement notice at 02.00 hours.

- Information-gathering phase, where an appraisal is made of what is happening and what needs to be done.
- This leads to a view being formed of who is at risk and what the risks might be.
- In general terms, the need is to make a continual assessment of the situation.

What are your considerations and do you have alternatives?

- There are standard operating procedures for this type of activity.
- The assessment is whether the standard ways of working will achieve the objectives in hand, or whether changes are required to enable the EHO

to continue to be safe in the particular circumstances of the specific activity.

- This risk-based methodology encourages the experienced EHO to take a systematic approach to their procedures using their knowledge, professional experience and intuition to assess the situation, and to consider possible ways of working to achieve the required outcome efficiently and safely.

The premises are on the top floor and it appears that the party has spilled onto the landing and the gathering gets a bit rowdy.

- This element of the model provides a pragmatic basis for checking excessive risk-taking behaviour.
- This stage of the model introduces a control on the individual who may, in the heat of the moment, become inclined to place themselves at unnecessary risk.

What do you do? Do you go in, make contact with base, and ask for assistance?

- Sometimes you may have to change something in the field as an additional control (your PRA should allow for this).
- How you implement those controls while the activity is actually happening, and the type of contingency in place in case it fails, should have been considered beforehand.

What do you do if an unexpected event occurs?

- A serious fight suddenly breaks out between two groups of people outside the property, blocking the EHO's exit. The home-owner (where the party is taking place) urges the EHO to enter the house to get away from the disturbance, but he/she is also concerned that violence could also break out inside the house at any time.

These considerations are key in terms of a rapid response when field workers are facing dynamic situations. However, please be mindful that additional risk controls implemented at the time of an assessment of risk in dynamic situations should not be considered as policy setting.

Select a system of work and continually assess until the activity is concluded.

- As the incident develops, re-evaluate the situation, tasks and persons at risk. Apply the above process to take account of any new hazards and introduce control measures as necessary to allow existing or new tasks to proceed.

Management questions

1   How does your organisation prepare your workers to manage risk in the field?
2   How are you going to ensure, so far as is reasonably practicable, that your staff present the desired behaviours in dynamic situations?
3   Do your predictive risk assessments enable your workers to make effective decisions when dealing with novel situations?
4   How do you prepare before and during the activity, and what is your contingency if the situation changes?

Case study 7.2 Security training for door supervisors working within licensed retail and events settings – Maybo (1)

This case study considers a DRA model and scenario-based training approach for work-related violence, which is now commonplace in a number of work sectors.

Context

Maybo, the specialist consultancy in workplace conflict and violence, has been a pioneer of the application of DRA in this field. Since the late 1990s its trademark SAFER Model* and Scenario Based Learning methodology has been widely adopted across a number of work sectors, including retail, transport, local government and health and social care. This case study looks specifically at its application to the security function within licensed retail and events settings.

Work-related violence is a complex risk area, as many factors can influence its causes and outcomes, with each set of circumstances being different. Human factors play a substantial role as different personalities interact in emotively charged situations. Escalation can occur in an instant, sometimes as the result of a single word or 'look'.

It was for these reasons that Maybo recognised the value of DRA in dealing with violence risks and the importance of developing the abilities of workers in this regard.

DRA as part of a wider risk-reduction strategy

Complex risks such as violence require sophisticated solutions and the World Health Organisation recommends a multi-strategy approach based on the Public Health Model, comprised of proactive and reactive elements.

---

*The SAFER device is a registered trademark of Maybo Limited

Proactive or 'primary' strategies involve a thorough understanding of the nature and extent of the risks and a focus on their causes and preventive controls. Events and licensed venue operators can normally predict certain conflicts related to the clientele they attract, e.g. by theme and/or music type, and also likely flashpoints such as access and dispersal. They can also predict that alcohol and other substances will be prevalent and can increase risk to staff and customers.

Reactive strategies include 'secondary' controls such as positive inter-personal skills of door supervisors to defuse conflict and 'tertiary' controls for an emergency. In this role, tertiary controls may include physical interventions to break up a fight and/or eject individuals.

Whilst effective DRA is clearly vital to help assess risks and inform decisions in a heated and fast-moving conflict situation, it is by nature 'reactive'. Organisations can and must also attempt to predict risks of violence and undertake further 'planned' risk assessments. Sector guidance encourages events and licensed venue operators and security providers to risk-assess the event, the venue and security activities performed which carry inherent risks, such as ejection.

Creating a Model for DRA

Human factors clearly play a key part in the build-up and management of a risk situation, including the ability to assess, communicate and make decisions. Having established the high relevance of DRA within a violence-reduction and management strategy, Maybo set about creating a helpful model and effective training approach. The SAFER Model (Figure 7.1) provides a simple framework to help workers assess the threat and risk in a situation and to consider and evaluate their options in dealing with it. The SAFER Model links to the 'POP' (see below) threat assessment model that was first used in law enforcement training in the United States and later adopted in the UK. Maybo added 'Situation' to cover the circumstances of the incident and key factors such as the role and expectations of those involved.

Step back: Do not rush in – physically or at least mentally step back and take in the situation on approach.

Assess threat: Consider potential threat and risks relating to **P**erson–**O**bjects–**P**lace–**S**ituation.

Find help: Tell someone what is happening and consider if help is needed.

Evaluation Options: Consider the choices available to you.

Respond: Choose the most appropriate option and continue to assess the situation and re-evaluate your strategy as necessary.

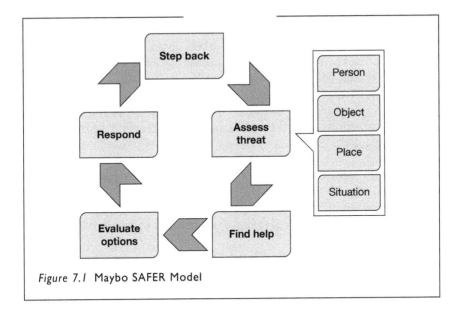

Figure 7.1 Maybo SAFER Model

Developing knowledge and skills: can DRA be learned?

Although much conflict and violence can be anticipated, risk assessed, planned for and prevented, its unpredictable nature means that there will always be a need for operational workers to 'think on their feet' as a situation unfolds.

Developing a model was relatively straightforward; the key question was whether training could develop an employee's ability to assess risk in a situation that is dynamic and unfolding. This reflected the 'nature or nurture' argument and it was Maybo's view that both elements played a part and, whilst some individuals were more adept compared to others of similar background and experience, all could improve from their start point.

Learning from experience can be painful, especially in this context, and some people are better than others at learning for themselves. Training in DRA can accelerate learning and:

- provide individuals with a structured method for assessing threat and risk in dynamic situations
- develop 'conscious competence'
- help reduce complacency and develop confidence
- increase self-awareness or 'mindfulness'
- improve observation skills
- challenge assumptions and stereotyping
- provide a framework to help objectively explain and justify decisions and actions.

How Maybo developed and delivers training in DRA

Maybo believed that for training in DRA to be effective it needed to be delivered in the specific context of the workers involved, i.e. directly relevant to the activities, environments and situations they experienced.

To achieve this, Maybo developed the SAFER scenario, a scenario-based learning approach whereby learners walk through a staged scene and share their observations and assessments. The scene develops in response to the decisions and actions they take and learning is facilitated by the trainer. This approach requires:

- thorough research and understanding of the job performed and risks faced
- identification of highly relevant risk scenarios for training
- clarity as to organisational policy, good practice and desired behaviours.

Door supervisor training utilises SAFER scenarios such as finding an individual slumped in a stairwell or toilets. The scenarios highlight the ease with which we make assumptions and the dangers that this presents.

Refreshing and building DRA

A one-off training event will raise interest and awareness levels for a time, but these can soon wane and complacency soon sets in. To maintain awareness and build knowledge and skills, an on-going blended approach to learning can be effective. Blended approaches help us to maximise learning and transfer through self-study and workplace engagement pre and post course-based training and can include:

- course-based training for developing understanding and skills through either external or 'in house' trainers
- e-learning for awareness and underpinning knowledge
- workplace coaches who help to review learning from incidents and to problem-solve
- workplace rehearsal of incident procedures and responses.

Risk scenarios can also be delivered in case study format where learners work in groups to apply the SAFER and POPS models, share observations and discuss responses. This does not have the same initial impact of a practical SAFER Scenario but the two approaches complement each other and case studies are particularly well suited to refresher training.

Establishment of DRA and scenario-based training within the UK private security industry

The private security industry is regulated by the Security Industry Authority (SIA), which introduced licensing of door supervisors in 2003. The Licence to Practise includes a comprehensive competency requirement delivered through course-based training, leading to a national qualification.

Maybo influenced the adoption of both DRA and scenario-based training as an SIA requirement in the core competencies for door supervisors and in the supporting training delivery guidance. It also helped SIA-endorsed Qualification Awarding Organisations to integrate DRA into their conflict management qualifications and the subsequent physical intervention training competency requirement.

Achieving effective and safe risk-based outcomes

Whilst well over 100,000 door supervisors have been introduced to DRA and it has been integrated into core training and qualifications, it is hard to evaluate its true impact. Part of the difficulty is poor recording and analysis of incidents and patterns, which is unsurprising, as door supervisors are typically part time and self-employed, often moving between venues. Where assaults and/or the use of force by door supervisors are known to have been reduced, this could also be down to a range of variables and not just their assessment; for example, improved control measures and training in alternatives to intervention.

Qualitative feedback and case studies have been easier to obtain and Maybo has found that learner feedback both on courses and in post-training (workplace) evaluations rates both DRA and scenario-based training highly. Learners frequently comment on its high relevance and meaning for their work, their increased levels of risk awareness and being better able to communicate and account for their decisions and actions.

The quality of the training design and its delivery is a factor in its effectiveness, and much rests on its transfer to the workplace and maintenance. Maybo favours a four-stage training approach that allows an evaluation to Kirkpatrick Level 4:

1   *Fit for Purpose:* Directly relevant to risks and needs, i.e. policy, context, behaviours experienced and activities undertaken
2   *Delivered as intended:* Highly relevant scenarios, professionally delivered and facilitated to draw on the experience which learners bring to training
3   *Transferred to the workplace:* Integrated into working practices, with supervision and monitoring to support positive behaviour change
4   *Maintained:* Refreshed, rehearsed, evaluated and embedded into practice to contribute to safer outcomes.

Figure 7.2 Door supervisors
Source: Phoenix Security

Maybo's experience is consistent with research and best practice in the field of violence reduction, in that training will contribute towards positive outcomes if it forms part of an integrated, multi-strategy, organisational approach. DRA is now a key element of training across most sectors experiencing conflict, challenging behaviours and violence.

TIP

Organisations should ensure that managers and staff are aware of possible cognitive biases which they bring to the training environment and when assessing and evaluating PRA and DRA.

Example: An EHO under-estimates a situation because he perceives it to be routine and does not adequately assess the escalating risk, missing vital clues.

Solution: Incorporate an understanding of awareness in risk perception into training and discuss ways to control this.

TIP

Staff involved in the risk-based decision-making line, e.g. back-room staff, should be targeted to attend DRA training and should have a basic understanding of the principles of risk assessment.

Example: Front-line staff may feel that they require additional controls to be implemented, such as an additional member of staff as back-up to ensure that their duties are carried out safely and effectively. However, back-room staff may not perceive the risk as sufficiently high to sanction the deployment of additional human resources.

Solution: In training courses use scenarios that include both front-line staff and back-room staff who support risk-based decision making, to highlight the importance of a shared understanding of risk and to enable the front-line worker to feel empowered to make decisions in the field that are supported.

TIP

Organisations that have developed partnerships through contracts and multi-agency working should ensure that cross-cutting risks are mutually understood and communicated, so as to enable the achievement of shared objectives.

Example: A couple have separated and their child lives with his mother and her new partner. Social Services has recorded the child's neglect on its records for over two years. The police have also been called to the home for minor disturbances. The child's attendance at school is sporadic and the school has raised concerns for the child's welfare with the parent and, more recently, with Social Services.

Solution: In multi-agency working situations organisations should set up a memorandum of understanding outlining where responsibilities begin and end, also how and when risk information can be shared and providing for an agreed method of on-going review to ensure its effectiveness.

TIP

Managers must lead by example and ensure that they have the necessary competence to manage risks at PRA and DRA level as an integral part of delivering their service.

Example: A manager working for a large company has completed an activity-based risk assessment for a member of his team whose work involves regular overseas work in countries noted for having civil unrest. The manager has not received training in PRA or DRA, and is not fully aware of the key hazards that his member of staff will encounter and how to effectively manage them.

Solution: Organisations must place occupational health and safety (OHS) as an essential business risk, along with other core functions, and ensure that understanding OHS responsibilities, such as risk assessment, is viewed as an important category of management skills and is made an integral part of the management appraisal/review process.

TIP

Consider human factors in terms of expected behaviours and addressing cognitive dissonance when developing your training course and facilitating learning.

Example: Workers bring to their role individual biases on risk perception and may not perceive what they do as being excessive. A security guard is called to the front desk of the local authority's department that deals with homelessness. The security guard has been told of a disturbance between two people and he wades in and inappropriately manhandles a woman, almost carrying her off the premises.

Solution: Training and facilitated learning in the application of DRA should be targeted, and given to line managers and not just to front-line staff in the field, as line managers have the responsibility for PRA.

Chapter summary

An opportunity to check your knowledge before you move on to the next chapter.

1　What does competence mean for the purposes of DRA?
2　What should organisations consider when considering procuring external consultants or training providers?
3　What is groupthink bias and how could this be a risk in an organisation?
4　What are the benefits of incorporating PRA and DRA training as a learning objective?
5　Why is it important that line managers (including those in the decision-making chain) and relevant back-room staff are involved in PRA/DRA training?
6　What areas of human factors would you include in the PRA and DRA training course for your organisation?
7　What are the issues to be considered when societal expectations and employer expectations are at odds with regard to delivering services?
8　What is the value of aligning your DRA and PRA training with organisational values and goals?

Chapter 8

Integrating learning from other organisations to improve performance

Learn from the mistakes of others.
You can't live long enough to make them all yourself.

(Eleanor Roosevelt)

Introduction

In Chapter 7, we looked at ways in which the three levels of risk assessment can be integrated within general risk assessment. In this final chapter we will explore the values of feedback and debriefing and how this closes the loop for continuous improvement referred to in Figure 4.1. We will also look at how you can learn from organisations that utilise dynamic risk assessment (DRA), in order to improve performance.

The topics that will be covered include:

- learning from other organisations to improve performance
- the value of debriefing and feedback
- Maybo case study (2)
- challenges of DRA implementation and conclusion.

Integrating learning to improve performance

Shropshire Fire and Rescue Service has developed its risk-assessment approach in addition to utilising DRA – which provides the Commander with a declaration of the tactical mode of operation; it has introduced analytical risk assessment (ARA). This more detailed assessment is undertaken at certain operational incidents. The ARA will then form the basis of a more detailed assessment. ARA is specifically applicable to the Fire and Rescue Service and is not generally utilised across other sectors. The approach developed by the Fire and Rescue Service for managing operational incidents, functions on the following three layers of risk assessment:

- generic risk assessment
- dynamic risk assessment
- analytical risk assessment.

The ARA introduces a third accepted stage in the risk-assessment process. Based upon the Brigade Order of Operation, the criteria on which ARA will be applied within the Fire and Rescue Service will be those situations where there is a need for four or more pumping appliances. In such situations, an ARA will be at the discretion of the 'Level Two' Tactical Commander. (Shropshire Fire and Rescue Service, 2011).

A summary of the key elements of ARA includes:

- a formalised assessment of the hazards, who or what is at risk from those hazards, likelihood and severity of risk
- a more detailed assessment of the existing control measures, with additional control measures introduced when the hazard or degree of risk requires
- confirmation that the DRA and tactical mode was/is correct
- inform on-going DRA process.

For further information on the principal fire-fighter data see Table 4.1, page 74.

In a presentation at a national health and safety conference Simon Pilling (West Yorkshire Fire and Rescue Service), when considering the reliability of DRA, said:

> Is it reliable? It's the best we've got!
> We recognize its limitations, therefore:
>
> - If in doubt, we default to defensive!
> - We always *back it up* with an analytical review and we keep reviewing.
> - We underpin it with evidence of being competent decision makers.
> (West Yorkshire Fire and Rescue Service
> Incident Command System (WYICS, 2002))

In Chapter 1 we discussed the challenges which emergency services face, by necessity operating on occasions in extremely hazardous environments. They have to balance achieving operational effectiveness with complying with the law. The HSE (2010) recognised the challenges faced by both the Fire and Rescue Service and the Police Service to achieve these health and safety duties within their operational work. For that reason, the HSE collaborated with both the Police Service and Fire and Rescue Service to produce two separate sets of high-level guidance.

- *Striking the balance between operational and health and safety duties in the Police Service*, www.hse.gov.uk/services/police/duties.pdf.
- *Striking the balance between operational and health and safety duties in the Fire and Rescue Service*, www.hse.gov.uk/services/fire/duties.pdf.

A pertinent question is, how can organisations successfully adopt DRA? There is no straightforward answer, but we attempt a simple answer to this in Chapter 4.

It is well known that many of the accidents and incidents of ill-health at work are a result of management failure and many of the case studies that we have covered in this book reflect this. How well does your organisation learn from past events?

If we follow the maxim that the best predictor of our future behaviour is our past behaviour, then why do some organisations find it hard to learn and apply lessons from past incidents to prevent a recurring theme?

It is human nature for people to focus on simple problems that appear to be clearly defined and where some knowledge or previous experience can assist, rather than on problems that are complex (Weick, Hopthrow, Abrams & Taylor-Gooby 2012). This is where having an effective risk management system can help to prioritise and deal with emerging risks and mitigate against the inertia of dealing with complex or difficult problems by leaving them in the 'too difficult box'.

Near-misses should be a vital part of learning lessons and reviewing health and safety performance. However, cognitive biases may affect how and whether near-misses are recorded. When reviewing past incidents or events people often overstate their ability to predict the outcome (hindsight bias). People either may try to justify their decisions or may genuinely believe that they selected a course of action because they knew what the outcome would be. However, at the time of the incident they may not have been sure of the outcome, and in fact it may have been more luck than judgement that the outcome was positive. Another issue with hindsight bias is when people do not recognise that the outcome was the result of luck and not knowledge or expertise. This may give their colleagues and themselves a false belief in their abilities. If people are making decisions without fully considering the risks, they may not have 'luck' on their side during the next incident. It is important to look realistically at how decisions were reached and to consider some of the cognitive biases that may have occurred.

In the authors' opinion the opportunity to learn lessons from incidents diminishes when investigations take an inordinate amount of time to complete, and this limits the learning opportunity for the organisation. This may be further exacerbated if the organisation has to deal with multiple incidents. Organisations need to ensure that investigations and learning are completed and shared with the relevant people within a reasonable time-frame. Accident investigations can vary in complexity and may take many months, and in some cases years to complete, and can involve more than one organisation. Whilst the investigation is on-going it may become apparent that specific factors either caused or indirectly contributed to the accident. This information is often not released to the public. This may be due to fear of affecting the integrity of the investigation, or for fear of possible litigation. This delay can impact on revised safety measures being quickly put in place, for not only the organisation involved but also other organisations which operate in a similar environment. The immediate impact of an accident (including the importance of learning lessons) may fade over time and this problem can be exacerbated if key personnel leave the organisation. It is therefore important to ensure that corporate memory is retained in order to effect learning.

The value of debriefing and feedback

As discussed previously, most risks are predictable and this allows organisations to put control measures in place to mitigate those risks. However, some risks are not always easy to predict, and where an organisation (or its staff) is involved in unpredictable situations, then a documented feedback process allows the organisation to incorporate lessons learned into its PRA. Where necessary, this process may include escalation to the strategic level. Once a risk comes to light it becomes predictable, and should be considered as such. The model of DRA offers a basis for learning and a structure for debriefing of incidents and activities (Tissington & Flin, 2005). How you communicate, disseminate and feed back information is crucial to your organisation and this should take place at all levels of the organisation. This is a key component of The 3-Level Risk Management Model™ in Figure 4.1.

TIP

Remember to include an exchange of health and safety risk information across multiple agencies where this is appropriate to facilitate learning.

With regard to our risk awareness, Weick et al. (2012) highlight the need for accurate feedback. If feedback is flawed, for example, inaccurate, this will lead to faulty representations of the situation. Our personal experience of risk exposure will be influenced by the frequency of events, and this may affect our ability to attend to infrequent hazards. Managers must take the lead in ensuring that reporting arrangements are established to achieve accurate feedback. The introduction of reporting and recording systems should ideally be incorporated into existing processes, in line with your agreed management information capture. If you are also able to demonstrate that you have relevant records before, during and after events, this will provide your organisation with a greater ability to improve performance and to demonstrate organisational learning.

What do workers feed back on?

This is a pertinent question and it should be determined by the outcomes required. When feeding back information at the dynamic level of The 3-Level Risk Management Model™, managers should make a judgement as to the usefulness of the information to their organisation. DRA should not necessarily be about feeding back every issue to the organisation; the focus should be on new hazards or novel situations not previously encountered, or deviation from established practices. Your senior management would not want to know all the minutiae, but, where risk escalation is relevant, such as having a strategic bearing on resourcing, this type of information may be of value to feed back at the SRA level through your established risk-escalation processes. Alternatively, it may remain an operational issue at the PRA level that requires a review of standard operating procedures or training. Nevertheless, managers need to develop a system (ideally aligned to existing business reporting systems) that enables the capture of relevant risk information after dynamic situations have occurred and ensure that it remains fit for purpose as part of review.

The following is a summary of the information that managers should consider:

- arrange timely debriefing so that take it takes place as early as possible, whilst information is still fresh in peoples' minds
- provide clear guidelines for staff on what aspects are important to record and feed back to management and others, whilst remaining flexible to divergent thinking
- clarify in debriefing what has been disregarded, and why
- challenge assumptions
- arrange formal/informal and/or one-to-one meetings as appropriate, to find out what went well and what didn't, sharing experiences as necessary.
- record lessons learned in debriefings, so that there is opportunity for other people to learn from them
- consider the review of PRA and/or strategic risk decisions (in light of new and significant hazards not previously identified/known) and, where appropriate, update policies, procedures and guidance
- communicate and disseminate relevant risk information (in good time) across the organisation, passing it to the right people to be able to act (as appropriate), as well as to multiple agencies, contractors and shared workplaces

- incorporate (as appropriate) lessons learned into staff training and learning programmes
- ensure effective and proportionate maintenance of relevant records before, during and after the activity, as part of a wider objective of continual improvement
- clarify the benefits of feeding back risk information and evaluation of risk-assessment training to your organisation
- remember, as Albert Einstein once said, to 'count what counts, not what can be counted'.

TIP

You need to know how your organisational hierarchy works both culturally and in terms of information flow. This will make all the difference to whether the risk information that is escalated (no matter how important) will sink or swim.

Remember, workers have a responsibility to highlight to line management where they believe that their knowledge, skills and experience have been unable to deal adequately with a situation or attain a desired outcome. This will give the worker the opportunity in a supportive environment to explain to line management what decisions they took and why. It also provides the line manager with the opportunity to give constructive feedback to the worker.

There are no hard and fast rules regarding the length of feedback and debriefs. Whilst time is a key consideration, managers should also be informed by the scale, nature and complexity of the activity. As a rule of thumb, debriefs should be succinct, focused and include all relevant workers, which may need to include multi-agency workers.

Case study 8.1 Security – Maybo (2)

This case study looks at how organisations have developed and integrated DRA in emergency and non-emergency situations as a key part of their violence-reduction strategies.

DRA is well established in emergency services and in Chapter 7 we looked at its adoption by the private security industry. It is clearly a valuable tool for those performing incident response and enforcement activities, and especially for those working in the community who may know little about the people, environments or situations they are approaching.

However, DRA is of wider value beyond the emergency services, and especially where work-related violence can be an issue, whether verbal abuse or higher-level threats of harm. This case study shows how organisations have applied and adapted the Maybo SAFER approach to meet their unique needs and challenges and how, through this process, they have identified and addressed gaps in policy, guidance and working practices.

How DRA has been adapted to meet the needs of non-emergency organisations

Health and social care services are amongst the most exposed to challenging behaviour and violence, in part due to the complex needs and difficult circumstances of patients and service users. Care staff can have an advantage, however, in that they often know well the individuals whom they support. This allows them to undertake individual behavioural risk assessments and to assess the risks which individuals present to themselves and others. It enables services to put in place specific support plans for each patient/service user that guide staff in how to prevent and respond to challenging behaviours.

DRA still plays an important role, but as part of a broader, person-centred approach focusing on understanding the 'function' or motive of the behaviour presented, which is often the result of an unmet need or communication difficulty. In such services there is opportunity to actively engage the service user and their family/advocates in strategies and plans that relate to their care and to review and evaluate progress.

Work-related violence tends to be concerned with staff safety, yet in care settings patients and service users are also vulnerable to harm, for example, from abuse, self-harm and restraint by staff. It is important therefore to focus on outcomes that can include a reduction in restraints as well as of incidents and assaults.

Key learning from the successful implementation of DRA

In its work with organisations across different sectors, subject specialist Maybo has shown DRA to be a valuable tool that will contribute to safer outcomes, provided that it is:

- *Part of a multi-strategy organisational approach:* Violence is a complex risk area that requires a combination of proactive and reactive strategies aligned to clear outcomes.
- *One aspect of risk assessment and not a panacea:* There is a risk in over-relying on DRA and becoming constantly reactive to events. Violence may be complex, but it can still be predicted and mitigated.
- *Designed and delivered effectively:* Relevance is critical if training is to address operational needs and fully engage staff. A training needs analysis anchored to risk assessment and involving staff will identify key risk scenarios, which become the bedrock of well-facilitated training delivered in a realistic environment.

The process and disciplines involved in developing DRA and scenario-based learning have delivered unexpected value for some organisations through highlighting gaps and inconsistencies in policy and guidance. This was very much the case in 2002–3, when Maybo undertook a training needs analysis in association with UNISON and a number of Ambulance Trusts in response to high levels of assaults against emergency personnel and calls for personal protective equipment.

The research identified 'first contact' with a casualty as a key area of risk and Maybo developed a DRA scenario to respond to this. The process, however, raised fundamental questions as to policy and guidance on the issue of crews 'standing off' or withdrawing where their assessment indicated significant risks of violence. In London the subsequent training, clarity over policy and improvements to risk information from Ambulance Control/Despatch played a key role in reducing assaults by 40 per cent.

Facilitating learning from experiences:

Maybo finds the Experiential Learning Cycle (Kolb, 1984) invaluable in drawing learning from incidents and near-misses (Figure 8.1). It helps to identify and address root causes, as opposed to simply bouncing from one bad experience to the next.

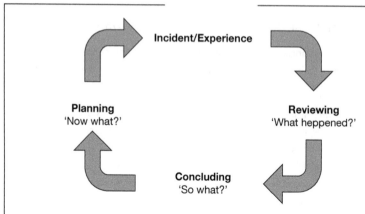

Figure 8.1 Maybo Cycle Model

This process can apply at every level, from the individual operative to the organisation, and inform risks, working practices, policy and controls such as training. It is important to focus on what is working well and reinforcing this, and not just to review things when they go wrong. In health, education and social care settings staff can help their service users to learn from experiences and to develop new skills and coping strategies.

Whilst we have a responsibility to investigate incidents and near-misses, how we do this is critical, and if we are to encourage staff to be open about issues and help us to resolve them we need to be supportive and not seek to lay blame. This can be a difficult balance, being sensitive and supportive to the needs of those involved who may be traumatised, whilst seeking to understand and learn from what happened so as to prevent a re-occurrence. Managers involved in reviewing incidents and facilitating support for individuals will benefit from training in how to establish immediate and on-going needs in a sensitive and effective way.

Maybo has found that the SAFER and POPS Models are helpful to staff when explaining and accounting for their decisions and actions to others, including managers or even the police, who were not present during the incident and may not appreciate how difficult it was to manage. Maybo also encourages organisations to regularly rehearse and test procedures and responses to incidents of challenging behaviour and violence – just as they do fire drills.

Challenges of DRA implementation

The success or otherwise of the application of DRA within your organisation will depend upon a number of interconnected factors that we have described throughout this book. Such as: senior management engagement and support, resources – which are not finite, the effective alignment of your health and safety management system to your organisational structures, the expected behaviour of workers.

Conclusion

In conclusion, organisations need to consider their own risk management arrangements. They need to avoid the temptation of planting another organisation's specific risk management methodology into their own company's risk management structures. What may be applicable to one organisation, for example, the financial environment, may not be in another, such as a nuclear plant. There are many organisations that have adopted DRA and can offer a sharing of good practice and learning. Some have been covered within this book. The key point is to look at DRA from an internal perspective, aligning organisational core values, service delivery and expected business outcomes within your organisational risk profile.

Remember, DRA is not effective as a stand-alone risk methodology. An organisation needs to ensure that it has an effective and integrated risk management system in place (Table 4.1), prior to adopting DRA. DRA can only be implemented successfully with senior management involvement and support.

TIP

Debriefing and feedback to the organisation(s) should be undertaken as early as possible, whilst information is fresh in the minds of those involved.

Example: A care worker visits a client and has been threatened verbally, but decides not to report the abuse to their line manager until two months later. This is only after someone else in the team has reported being assaulted by the same client.

Solution: Train managers involved in the decision-making process in the art of debriefing and feedback, which are management skills akin to facilitating team meetings, carrying out one-to-one meetings or appraisal meetings. This can also be incorporated within PRA/DRA training. Multi-agency working arrangements, wherever possible, should have in place agreements, memoranda of understanding (MoU) and/or contracts that detail occupational health and safety responsibilities concerning the sharing of information on health and safety risks.

TIP

The outcome of investigations following an incident and the valuable learning of lessons to improve performance should be shared with the right people, at the right time, to enable timely and effective change. This includes third parties, as this will help to build long-term trust, collaboration and efficiencies in service delivery.

Example: A parade takes place in a busy city centre and the expected number of spectators in the area is grossly under-estimated. Multi-agency planners such as the local authority, event organisers and local transport organisations have to deal with the fall-out from crowd congestion.

Solution: Share information from past events and use management information from current events to help plan and inform future and/or recurring events.

TIP

Learning from activities or incidents that have a significant effect on risk should be fed through established reporting systems to effect change in PRA and future delivery of training and learning.

Example: An investigation into a serious accident identified that the PRA had not been revised and training for the activity had not been updated following a previous incident.

Solution: Where possible, incorporate a case study of the event into training and learning programmes and ensure the capture of all relevant risk issues. As part of a sound investigation, the action plan resulting from recommendations should include the need to review the PRA (prompted by details of debrief and feedback) and, where necessary, update training accordingly. Additionally, the organisation's reporting systems should be checked to ensure that they remain fit for purpose and report forms/checklist should be revised as is necessary, ensuring that the reporting process has been adequately signed off.

TIP

Promote the culture of a learning- and solution-based organisation, rather than a 'blame culture', as part of continual improvement. This will enable workers to confidently report and feed back on risk issues, such as gaps or failings in risk control systems, following an activity or incident.

Example: An incident occurred (with no actual injury) where a worker withheld some information (due to fear of reprisals) concerning weaknesses in the safety management system, which led him to cut corners in his operation. The result was a failure to feed back vital information that would have enabled the organisation to learn, to make organisational improvements to its systems of work and, ultimately, to become more effective and efficient in its service delivery.

Solution: Educate managers (involved in the decision-making chain) to lead by example and to underpin the organisation's core values through action. Encourage learning and sharing of information, and tackle herd behaviour (when conformity stifles the ability to highlight risk issues), as part of the wider organisational commitment to continual improvement.

TIP

The type and nature of information that is collected at the debrief and fed back to management should be determined by the required organisational outcomes for the activity.

Example: Two organisations are working together, one of which has adopted DRA methodology within its risk management system. The other, which has not, fails to share crucial information to the multi-agency organisation following an internal investigation resulting from a major incident (which results in enforcement action), which identifies where potential learning could both achieve improvement across organisations and mitigate occupational health and safety risks. The organisation is fearful that, by sharing the information, it may prejudice its standing in court, and decides to withhold key information until after the court case has concluded (at least two years later).

Solution: Develop structured processes to enable both formal and informal feedback from operations, be this through debriefs directly after the event or as close as possible to its conclusion, or during team meetings or one-to-one sessions, which should be managed carefully.

Chapter summary

In the final chapter you have an opportunity to check your knowledge against the area covered.

1 Why is it important to share risk information with multi-agency organisations?
2 What does ARA stand for and which organisation uses this to inform its risk-based decision making?
3 What are the benefits of feedback and debriefings?
4 Why is it important to provide feedback and debriefings as early after an event as possible?
5 Who is responsible for ensuring that the information on new risks is shared and escalated across teams and/or the organisation?
6 What are the methods to communicate risk information?
7 Why do organisations find it challenging to learn from incidents?
8 How can organisations learn quickly from repetitive failings?
9 What organisational factors should be considered in management information capture following feedback and debriefings, and why?
10 How can organisations measure success, what does it look like and who sets and measures its effectiveness?

Appendix: Professional organisations related to health and safety and risk management

ALARM	National Forum for Risk Management in the Public Sector *www.alarm-uk.org*
ASSE	American Society of Safety Engineers *www.asse.org*
BCI	The Business Continuity Institute *http://thebci.org/*
BSC	British Safety Council *https://www.britsafe.org/*
CIH	Chartered Institute of Housing *www.cih.org*
DfES	Department for Education Services *www.education.gov.uk/*
EPS	Emergency Planning Society *https://www.the-eps.org/*
IFE	The Institution of Fire Engineers *www.ife.org.uk/*
IIRSM	International Institute of Risk and Safety Management *www.iirsm.org*
IoD	Institute of Directors *www.iod.com*
IOSH	Institution of Occupational Safety and Health *www.iosh.co.uk*
IRM	The Institute of Risk Management *www.theirm.org/*
ISMA	International Stress Management Association *www.isma.org.uk*
LHSG	The London Health & Safety Group *http://londonhealthandsafetygroup.org/*
NCVO	National Council for Voluntary Organisations *www.ncvo-vol.org.uk*
RoSPA	The Royal Society for the Prevention of Accidents *www.rospa.com*
RSPH	Royal Society for Public Health *www.rsph.org.uk*

Glossary of acronyms

ALARP	as low as reasonably practicable
ARA	analytical risk assessment
CDM	classical decision making
CPD	continuing professional development
CSR	corporate social responsibility
DRA	dynamic risk assessment
ERIC	Eliminate, Reduce, Isolate, Control
EU	European Union
EWS	early warning system
HSE	Health and Safety Executive
HSE-MS	health and safety management system
HSWA	Health and Safety at Work, etc Act (also HASAWA)
ILO	International Labour Organisation
ISO	International Standards Organisation
KPI	key performance indicator
MEOWS	Modified Early Obstetric Warning System
MEWS	Modified Early Warning System
MHSWR	Management of Health and Safety at Work Regulations
MOC	management of change
MSS	management system standard
NDM	naturalistic decision making
NEWS	National EarlyWarning Scoring (System)
NHS	National Health Service
OH	occupational health
PC	Police Constable
PDCA	Plan, Do, Check, Act
PEST	Political, Economic, Societal, Technical
PHSA	Police (Health and Safety) Act
PPE	personal protective equipment
PRA	predictive risk assessment
RAM	risk assessment matrix
RCP	Royal College of Physicians

RPDM	recognition-primed decision making
SFAIRP	so far as is reasonably practicable (also SFRP)
SHEQ	safety, health, environment, quality (also HSEQ or QEHS)
SKATE	Skills, Knowledge, Attitude, Training, Experience (= Competence)
SMART	Specific, Measurable, Achievable, Right (or Realistic), Timely
SRA	strategic risk assessment
TRA	Theory of Reasoned Action
UK	United Kingdom
UN	United Nations
US / USA	United States / United States of America

Glossary of dynamic risk assessment-related language

Actual incident A specific event or extended condition that resulted in a significant, unwanted and unintended impact on the safety or health of people, on property, on the environment, on legal/regulatory compliance or on security as it relates to operations integrity.

Audit A process which evaluates activities, facilities or processes against management system requirements (or expectations).

Business environment The current and future characteristics of the place where the organisation, its supply chain and competitors operate. Understanding the business environment sets the context for understanding the needs and expectations of stakeholders, and determines the necessary scope of the management system.

Business objectives Objectives set by management in the context of their business environment to meet the requirements of their stakeholders.

Competency The skills, knowledge, attitude, training and experience ('SKATE') required of personnel to successfully perform job-related tasks to achieve business objectives.

Contract A legally enforceable, documented agreement that includes purchase order and service agreement.

Controlled documents Documents managed by a system that assigns custodians who control modifications, circulation and distribution. The system ensures removal of obsolete or superseded material.

Culture 'The way we do things around here' (HSE, 1997).

Data Quantitative or qualitative information derived from testing and/or measurement.

Deployment The act of providing initial communication and training for a system, process, procedure or programme.

Document owner The individual assigned to provide management oversight and final approval of modifications, circulation and distribution of a document.

Documentation Electronic or hardcopy manual, procedure, drawing, legal agreement, correspondence, record or reference that contains information.

Dynamic risk assessment (DRA) A continuous mental assessment of the risk in a changing environment which informs decision making to provide an acceptable level of safety.

Employee Any individual on the payroll of the organisation engaged in activities for which the individual is paid.

Expectations The minimum requirements set out in a documented management system standard.

External audit An assurance process conducted by an audit team from outside the organisation which determines the extent to which operations comply with the expected requirements of a specified management system. Synonymous with 'Third party audit'.

Facilities Physical equipment and/or plant, including large mobile equipment, involved in the performance of operations.

Function A specialised organisation that supports the operating organisation. Often used to describe that part of a corporate function specifically assigned to work within an asset.

Gap An identified deficiency in addressing and/or complying with a management system expectation or other defined requirement.

Gap analysis A method or means to identify deficiencies between management system expectations and actual practices. The results of such analyses are documented.

Governance The act, process or power of governing.

Guidelines A second tier of requirements, developed by organisations or their regulators to further define expectations. Synonymous with 'Approved codes of practice'.

Hazard A potential source of serious harm to people, property or the environment.

Higher risk The combinations of probability and consequence on a risk matrix that constitute an area of higher risk, as defined by management.

Human factors The integration and application of scientific knowledge about people, facilities and management systems to improve their interaction in the workplace, in order to achieve improved safety, health, environmental and financial performance. Human factors considerations may include: workplace design; equipment design; work environment; physical activities; job design; information transfer; and personal factors.

Implementation A multi-step process involving 1) allocation of resources, roles, responsibilities and authorities, 2) competency, training and awareness, 3) consultation, participation and communication, 4) documentation and control of documents and 5) system stewardship for effectiveness and continual improvement. Sections 4.4.1–4.4.5 of ISO 14001 and OHSAS 18001 refer to implementation.

Incident A specific event or extended condition that resulted (i.e. actual incident) or could have resulted (i.e. near-miss incident) in a significant, unwanted and unintended impact on the safety or health of people, on property, on the environment, on legal/regulatory compliance or on security as it relates to operations integrity. See *Actual incident* and *Near miss incident*.

Initial training The training required before assuming job duties for a new hire, prior to performing the duties of a new position for an existing employee, or that acquired during the initial performance of job duties.

Interface A place or time at which independent systems and/or operations of two or more companies, functions, assets, departments or groups of works should act with or communicate with each other. With respect to HSE management systems, the term 'interfaces between operations' refers to two types of interfaces: 1) Systems-level interface between two or more companies, functions or assets that requires system documentation (e.g. bridging documents, service agreements, etc.); 2) Procedural-level interfaces, typically occurring at sites with shift operations or simultaneous work activities that require special procedures (e.g. shift-change procedures, simultaneous operations procedures or manuals).

Internal audit A process conducted by members of the assessed organisation, operating independently, which determines the extent to which operations comply with the expected requirements of a specified management system. Synonymous with 'Second party audit'.

Key positions Those positions which are part of an organisation and which could have a significant impact on operations integrity through individual decision-making authority, operational control or individual actions. NB: These positions are sometimes held by third parties.

Leadership The action of leading a group of people or an organisation.

Line or line management Management or individuals responsible for the day-to-day functioning of an organisation.

Major project A project requiring senior management endorsement.

Management Senior management are typically corporate function managers to whom operations and department managers report. Senior operations management are typically the department level of management directly responsible for operations. Operations management are typically operating organisation personnel responsible for direct supervision of operations. Site management are typically the most senior level of supervision working on site.

Near-miss incident An undesirable event which could have resulted, under slightly different circumstances, in an unwanted impact on the safety or health of people, property or the environment.

Occupational health programme All activities addressing workplace health hazards and employee health. It includes identification, evaluation and control of health hazards, monitoring of worker exposure, communication of health hazard knowledge, determination of employees' medical fitness for duty and providing or arranging for medical services necessary for the treatment of occupational illnesses or injuries.

On-going training Training oriented toward improving an employee's performance in a given position. It could include training to keep the employee current with changing technology or to provide improved skills.

Operating organisation A subset of a corporate function that is responsible for activity involving the production, manufacture, use, storage, movement of materials or utilisation of resources to produce an output.

Operation Any activity involving the production, manufacture, use, storage or movement of material. Also, the utilisation of resources by a function to produce an output.

Operations integrity Operations integrity addresses all aspects of the organisation's business, including security, which can impact on safety, health and environmental performance.

Performance measures Indicators used to determine if a system or groups of systems are meeting or progressing toward expected objectives and are likely to continue the progress. Included are two types of measures: 1) Active measures, utilised to measure the degree to which the execution of a system conforms to requirements (i.e. leading indicators); 2) Reactive measures, utilised to measure the degree to which system objectives are being met (i.e. lagging indicators).

Personnel A broad term generally used to describe employees or contractors (i.e. third parties) or other authorised individuals involved in a specific activity. Synonymous with 'Workforce'.

Practice Approved method or means of accomplishing stated tasks.

Predictive risk assessment (PRA) The planned process for determining and documenting risks related to recognised hazards and known situations, and the identification of the appropriate means which, when communicated and implemented, eliminate or control the risk to tolerable levels.

Procedure A documented series of steps to be carried out in a logical order for a defined operation or in a given situation.

Process A series of actions, changes or functions that bring about an end or result.

Product A material that has been manufactured, refined or treated and is sold or exchanged.

Programme A plan that includes scheduled events or activities.

Project A planned undertaking with a specific objective and defined scope. Includes new constructions and additions or revisions to existing facilities.

Qualified Possessing knowledge and skills, through training and experience, required to successfully perform job-related tasks.

Quality The ability of a product, service or activity to meet or exceed requirements.

Quality assurance The planned and systematic actions necessary to ensure that a product, service or activity is able to meet or exceed requirements.

Quality control A control that determines whether or not a product, service or activity meets requirements through measurement, testing and/or inspection.

Records Documentation that reflects evidence of past events or actions (e.g. inspections, incident investigations, training, maintenance performed, air emissions, reports such as risk assessments and HAZOP (hazard and operability study), etc.)

Red mist A feeling of extreme competitiveness or anger that temporarily clouds one's judgement.

Refresher training Training that reinforces employees' skills and understanding, enabling them to continue to perform safely, correctly and in an environmentally sound manner.

Risk The effect of uncertainty on objectives. A function of the probability (or likelihood) of an unwanted incident, combined with the severity of its potential consequences.

Risk analysis The development of a qualitative or quantitative estimate of risk.

Risk appetite An expression of risk tolerance, often described in the values and beliefs of the organisation.

Risk assessment The process of judging the significance of risk and determining whether further risk reduction is warranted.

Risk assessment matrix (RAM) A standard matrix on which the levels of probability and the severity of consequences are defined is usually contained in an organisation's management system manuals or similar documentation. Plotting incident scenarios on this matrix is a means of prioritising and communicating the risk level associated with assessed activities.

Risk management The application of policies and practices to the tasks of identifying, assessing and controlling risks in order to protect human life, the environment, physical assets and company reputation in a cost-effective manner.

Root cause(s) The most fundamental reason (or reasons) for the occurrence of an event that, when corrected, will prevent (or significantly reduce the probability or consequences of) its recurrence.

Security A set of controls or measures taken to protect personnel from injury and to protect physical assets from damage or loss as it relates to HSE, including where this could have an adverse environmental impact.

Site The place where something was, is or will be located.

Stakeholders Those interested (holding a stake) in the achievement of business objectives. We categorise these in five main groups: 1) shareholders (or investors), 2) employees, 3) suppliers (those with whom we do business), 4) customers, 5) society (including regulators, pressure groups, media, and the public).

Standard A documented level of requirement which must be achieved or is a target for achievement.

Stewardship The process of being accountable for an activity or function and its performance against a goal, including maintaining knowledge of its status and reporting its condition to higher management.

Strategic risk assessment (SRA) The high-level risk-based framework to enable proportional risk controls, resources and priorities to deliver organisational objectives in service delivery.

System A structured means of ensuring that business objectives are achieved and sustained. The five characteristics of current ISO HSE management systems are: 1) Policy, 2) Planning, 3) Implementation and operation, 4) Checking and 5) Management review. These characteristics will likely be amended at their next revision (see 'Management system standards and BS OHSAS 18001' in Chapter 2) as a result of the issue of *Annex SL* in April 2012, as follows: 1) Context of the organisation, 2) Leadership, 3) Planning, 4) Support, 5) Operation, 6) Performance evaluation and 7) Improvement.

Third party A contractor, subcontractor, supplier or vendor providing materials or services in accordance with specifications, terms and conditions documented by a contract agreement and signed by both parties.

Values and beliefs Operating philosophies or principles that guide an organisation's internal conduct as well as its relationship with the external world.

Workers/workforce A broad term generally used to describe employees or contractors (i.e., third-parties) or other authorised individuals involved in a specific activity. Synonymous with 'Personnel'.

References

ACPO (Association of Chief Police Officers) (2009). *The Management of Police Pursuits – Guidance*. Association of Chief Police Officers of England, Wales and Northern Ireland.

ACPO (2011). *The National Decision Model*, ACPO, www.acpo.police.uk/documents/president/201201PBANDM.pdf, accessed 22/11/2013.

Ajzen, I. (1991). The theory of planned behavior. *Organizational Behavior and Human Decision Processes*, 50, pp 179–211.

Alpert, G.P. & Dunham, R.G. (1990). *Controlling Responses to Emergency Situations*. US Department of Transportation National Highway Traffic Safety Administration.

Asbury, S.W. (2013). *Health and Safety, Environment and Quality Audits – A Risk-based Approach*. Routledge.

Asbury, S.W. & Ball, R. (2009). *Do the Right Thing – The Practical, Jargon-free Guide to Corporate Social Responsibility*. The Institution of Occupational Safety and Health.

Asch, S.E. (1940). Studies in the principles of judgments and attitudes: II. Determination of judgments by group and by ego-standards. *Journal of Social Psychology*, 12, pp 433–465.

Asch, S.E. (1948). The doctrine of suggestion, prestige and imitation in social psychology. *Psychological Review*, 55, pp 250–276.

Atkinson, J.W. (1957). Motivational determinants of risk-taking behaviour. *Psychological Review*, 64(6, Pt.1), pp 359–372

Bargh, J.A. & Chartrand, T.L. (1999). The unbearable automaticity of being. *American Psychologist*, 54(7), pp 462–479.

Baron, J., Badgio, P. & Gaskins, I. W. (1986). Cognitive style and its improvement: A normative approach. In R.J. Sternberg (ed.), *Advances in the Psychology of Human Intelligence*, Vol 3, pp 173–220. Erlbaum.

BBC (1997). *Spiral to Disaster* (video) BBC Active, www.bbc.co.uk/programmes/p00fncvs, accessed 02/02/1998.

Bernstein, P.L. (1996). *Against the Gods – the Remarkable Story of Risk*. Wiley.

Boyle, T. (2002). *Health and Safety: Risk Management*. The Institution of Occupational Safety and Health.

Brehmer, B. (1992). Dynamic decision making: human control of complex systems. *Acta Psychologica*, 81, pp 211–241.

BSI (British Standards Institution) (2007). BS OHSAS 18001:2007 *Occupational Health & Safety Management Systems – Specification*. British Standards Institution.

Clements, D. (2013). Interview with Stephen Asbury, Coalville, 1 July 2013.

Cohen, M.S., Freeman, J.T. & Thompson, B. (1998). Critical thinking skills in tactical decision making: A model and a training strategy. In J.A. Cannon-Bowers and E Salas (eds), *Making Decisions Under Stress: Implications for Individual and Team Training*, 155–190. American Psychological Association

Columbia Accident Investigation Board (2003). *Report Volume I.* National Aeronautics and Space Administration and the Government Printing Office.

Crichton, M. (2013). www.goodreads.com/quotes/188569-if-you-don-t-know-history-then-you-don-t-know-anything, accessed 25/9/2013.

Cullen, The Hon. Lord W. Douglas (1990). *The Public Inquiry into the Piper Alpha Disaster*. HM Stationery Office.

DCLG (Department for Communities and Local Government) (2013). *Fire and Rescue Authorities Health, Safety and Welfare Framework for the Operational Environment*. DCLG.

Dillon, R.L. & Tinsley, C.H. (2005). Whew that was close! How near miss events bias decision making. *Academy of Management Best Paper Proceedings*.

England and Wales High Court (Queen's Bench Division) Decisions (2012). Cornish Glennroy Blair-Ford v CRS Adventures Ltd [2012] EWHC 2360 (QB) (13 August 2012), www.bailii.org/ew/cases/EWHC/QB/2012/2360.html, accessed 09/01/2013.

Evans, J.St.B.T. & Over, D.E. (1996). *Rationality and Reasoning*, Psychology Press.

FBU (Fire Brigades Union) (2008). *In the Line of Duty – Firefighter Deaths in the UK since 1978*. Report by Labour Research Department for the Fire Brigades Union.

Federal Aviation Administration (1978). Lessons learned. United Airlines, Flight 173, MD DC-8-61, N8082U. http://lessonslearned.faa.gov/ll_main.cfm?TabID=1&LLID=42, accessed 21/11/2013.

Fielding, N. (1994). *Just Boys Doing Business: Men, Masculinities and Crime Cop Canteen Culture*. Routledge.

Financial Reporting Council (2005). Internal Control. Revised Guidance for Directors on the Combined Code (October), www.ecgi.org/codes/documents/frc_ic.pdf, accessed 15/04/2013.

Fire Service Circular 5/1995 and Scottish Fire Service Circular 4/1995 (1995). *Health and Safety in the Fire Service*. Home Departments.

Fire Service Manual (2008). *Fire Service Operations Incident Command*, 3rd edition, Volume 2. The Stationery Office.

Firenet (2013). Firefighters entering a building, http://fire.org.uk/forum/index.php?topic=5668.15, accessed 10/09/13.

Fishbein, M. & Ajzen, I. (1975). *Belief, Attitude, Intention and Behaviour: An Introduction to Theory and Research*. Addison-Wesley.

Flin, R. (1996). *Sitting in the Hot Seat – Leaders and Teams for Critical Incident Management*. Wiley & Sons.

Flin, R., O'Connor, P. & Crichton, M. (2008). *Safety at the Sharp End: A Guide to Non-technical Skills*, Ashgate.

FSEB (Fire Services Examinations Board) (2003). Study Note LFFSEC4.1315.

Fuller, C.W. & Vassie, L.H. (2004). *Health and Safety Management – Principles and Best Practice*. Prentice Hall.

Gardner, D. (2009). *Risk*. Virgin Books.

Gutteling, J. & Wiegman, O. (1996). *Exploring Risk Communication,* Advances in natural and technological hazards research, v.8. Dordrecht and London: Kluwer Academic.

Harrison, D. (2004). Fatal stabbing of detective prompts body armour review, *Telegraph*, 23 May, www.telegraph.co.uk/news/uknews/1462600/Fatal-stabbing-of-detective-prompts-body-armour-review.html, accessed 23/05/2013.

Hillsborough Independent Panel (2012). *The Report of the Hillsborough Independent Panel 2012*, HC581. The Stationery Office.

HM Fire Service Inspectorate (1998). *Dynamic Management of Risk at Operational Incidents*. The Stationery Office.

HM Fire Service Inspectorate (2002). *Fire Service Manual. Volume 2: Fire Service Operations Incident Command*. The Stationery Office.

Home Office (1996). *Police Health and Safety*, vol. 2–4 *The Police Policy Directorate*. Home Office.

Home Office (1998). *Dynamic Management of Risk at Operational Incidents. A Fire Service Guide Pamphlet*. The Stationery Office.

Home Office Circular (1998). *Police Health and Safety*, HOC 27/98, 29 June. Home Office.

Horvath, P. & Zuckerman, M. (1993). Sensation seeking, risk appraisal and risky behavior. *Personality and Individual Differences*, 14, pp 41–52.

HSE (Health and Safety Executive) (1997). *Successful Health and Safety Management*, 2nd edition, HSG65. HSE Books.

HSE (2005). *A Review of Safety Culture and Safety Climate Literature for the Development of the Safety Culture Inspection Toolkit*. Research Report RR367. HSE Books.

HSE (2009a) *Reducing Error and Influencing Behaviour*, HSG48, 2nd edition. HSE Books.

HSE (2010). Striking the balance between operational and health and safety duties in the Fire and Rescue Service, www.hse.gov.uk/services/fire/duties.pdf, accessed 15/10/2012.

HSE (2011). Five steps to risk assessment, indg163(rev3) revised 06/2011. HSE Books.

HSE (2013a). Terms of reference for the Mythbusters challenge panel, www.hse.gov.uk/myth/myth-busting/terms-of-reference.htm, accessed 01/08/13.

HSE (2013b). From Mythbusters challenge panel decision number 197, www.hse.gov.uk/myth/myth-busting/2013/case197-school-sports-day.htm, accessed 01/08/13.

ILO (International Labour Organisation) (2001). Guidelines on occupational safety and health, ILO-OSH 2001, www.ilo.org/safework/info/standards-and-instruments/WCMS_107727/lang—en/index.htm, accessed 12/10/2012.

IoD (Institute of Directors) (2009). *Leading Health and Safety at Work*, INDG417. Institute of Directors and HSE, www.hse.gov.uk/pubns/indg417.pdf.

IRM (Institute of Risk Management) (2002). *A Risk Management Standard*. The Institute of Risk Management.

IRM (2009). *A Structured Approach to Enterprise Risk Management (ERM) and the Requirements of ISO 31000*. The Institute of Risk Management, The Association of Insurance and Risk Managers (AIRMIC) and The Public Risk Management Association (ALARM).

ISO (International Standards Organization) (2009). *Risk Management – Principles and Guidelines*, ISO 31000:2009 and *Risk Management Vocabulary*, ISO Guide 73. International Standards Organization.

ISO (2012). *Annex SL* (previously *ISO Guide 83*) of the *Consolidated ISO Supplement of the ISO/IEC Directives*. International Standards Organization.

Jacobs, E. (2010a). Vulnerable workers – take care on the hoof, *Safety and Health Practitioner*, www.shponline.co.uk/features/features/full/vulnerable-workers-take-care-on-the-hoof, accessed 03/12/2012.

Jacobs, E. (2010b). In proceedings of *Lone Worker Safety Conference*, 11 May 2010, Earls Court, London.

Jacobs, E. (2013). In proceedings Making Dynamic Risk Assessment Part of an Effective Risk Management Plan, *Health and Well-being at Work*, 5 March 2013, National Exhibition Centre, Birmingham.

Janis, I.L. & Mann, L. (1977). *Decision Making: A Psychological Analysis of Conflict, Choice, and Commitment*. Free Press.

Janis, I.L. (1982). *Groupthink: Psychological Studies of Policy Decisions and Fiascoes*. Houghton Mifflin.

Joint Committee on Fire Brigade Operations (1996). *Guidance on the Application of Risk Assessment in the Fire Service*. Home Office.

Kemp, C., Norris, C. & Fielding, N.G. (1992). *Negotiating Nothing: Police Decision-making in Disputes*. Avebury.

Klein, G. (1998). *Sources of Power – How People Make Decisions*. MIT Press.

Klein, G. & Klinger, D. (1991). Naturalistic decision making. *Human Systems IAC Gateway*, XI, 3, 2, 1 (Winter), pp 16–19.

Kolb, D.A. (1984) *The Kolb Learning Cycle*, www.ldu.leeds.ac.uk/ldu/sddu_multimedia/kolb/kolb_flash.htm (accessed 22/11/2013).

Lipshitz, R., Klein, G., Orasanu, J. & Salas, E. (2001). Focus article: Taking stock of naturalistic decision making. *Journal of Behavioural Decision Making*, 14, pp 331–352.

Lusk, G.L. (2008). Assessing decision making in dynamic, high-risk environments to enhance Amarillo Fire Department safety, www.usfa.fema.gov/pdf/efop/efo41897.pdf, accessed 25/03/2013.

Mail (2008). www.dailymail.co.uk/news/article-507651/Coastguard-sacked-daring-clifftop-rescue-13-year-old-schoolgirl-boss-says-We-dont-want-dead-heroes.html, accessed 26/9/2013.

McBride, M. (1996). *Street Survival Skills – An Operational Guide to Police Officer Safety*. Police Review Publishing Co.

Metropolitan Police Authority (2004). *Aide Memoire. Safe Policing at High Risk Incidents*. MPS Health and Safety Branch.

Metropolitan Police Federation (2013). https://www.metfed.org.uk/metline?id=2067, accessed 27/9/2013.

Metropolitan Police Service (2010). Reply to freedom of information request, www.met.police.uk/foi/pdfs/disclosure_2010/december/2010110000129.pdf, accessed 25/9/2013.

Morris, W. (2013). www.manchestereveningnews.co.uk/news/greater-manchester-news/fire-service-in-clear-over-death-1126244, accessed 27/9/2013.

Murdermap (2013). www.murdermap.co.uk/pages/cases/case.asp?CID=235825746, accessed 27/9/2013.

NASA (2013). Commentary by NASA host during Stephen Asbury's visit to MOCR2, Johnson Space Center, Houston, 28 April.

National Aeronautics and Space Act (1958). http://history.nasa.gov/spaceact.html, accessed 22/11/2013.

Obama, M. (2013). www.goodreads.com/quotes/tag/decision-making, accessed 25/9/2013.

Oligny, M. (1994). Burnout in the police environment. *International Criminal Police Review*, 446, pp 22–25.

Ravasi, D. & Schultz, M. (2006). Responding to organizational identity threats: exploring the role of organisational culture. *Academy of Management Journal*, 49(3), pp 433–458.

RCP (Royal College of Physicians) (2012). *National Early Warning Score (NEWS): Standardising the Assessment of Acute Illness Severity in the NHS*. Report of a working party. Royal College of Physicians.

Ritchie, B. & Marshall, D. (1993). *Business Risk Management*. London: Chapman & Hall.

Roosevelt, E. (n.d.). www.goodreads.com/quotes/20393-learn-from-the-mistakes-of-others-you-can-t-live-long, accessed 2/10/2013.

Roosevelt, T. (n.d.). www.brainyquote.com/quotes/quotes/t/theodorero403358.html, accessed 25/9/2013.

Sefton, A. (1992). In Flin (1996: 23). *Sitting in the Hot Seat – Leaders and Teams for Critical Incident Management*. Wiley & Sons.

Sharot, T., Korn, C.W. & Dolan, R.J. (2011). How unrealistic optimism is maintained in the face of reality. *Nature Neuroscience*, 14, pp 1475–1479.

Shropshire Fire & Rescue Service (2011). Analytical Risk Assessment, Brigade Order 5 Part 5, Reference OPS5PT5, www.shropshirefire.gov.uk/sites/alpha.shropshirefire.gov.uk/files/brigade-orders/incident-command/operations-5-part-5-analytical-risk-assessment.pdf, accessed 05/01/2013.

Sloman, S.A. (1996). The empirical case for two systems of reasoning. *Psychological Bulletin*, 119, pp 3–22.

Slovic, P., Kunreuther, H. & White, G. (1974). Decision process rationality and adjustment to natural hazards. In G.F. White (ed.), *Natural Hazards*. Oxford University Press.

Soane, E. & Chmiel, N. (2005). Are risk preferences consistent? The influence of decision domain and personality. *Personality and Individual Differences*, 38, pp 1781–1791.

Stanovich, K.E. & West, R.F. (2000). Individual differences in reasoning: implications for the rationality debate. *Behavioral & Brain Sciences*, 23, pp 645–665.

Starr, C. (1969). Social benefits versus technological risks. *Science*, 165(3899) (19 September), pp 1232–1238.

Stasser, G. & Stewart, D. (1992). Discovery of hidden profiles by decision-making groups: solving a problem versus making a judgment. *Journal of Personality and Social Psychology*, 63, pp 426–434.

Stasser, G. & Titus, W. (2003). Hidden profiles: A brief history. *Psychological Inquiry*, 3–4, 302–311.

Stenhouse, C. (2013). Interview with Stephen Asbury, Burton Hospitals NHS Foundation Trust Head Office, Burton upon Trent, 7 August.

Sterling, J. (1972). *Changes in Role Concepts of Police Officers*. Professional Standards Division International Association of Chiefs of Police.

Stevenson, A. (2013). Interview with Stephen Asbury, Sahara Force India HQ at Silverstone, 22 July.

Stewart, E. (1997). *Improving Decision Making in Emergency Situations*. Police Research Award Scheme. Grampian Police Home Office Police Department.

Sullenberger, C.S. with Zaslow, J. (2009). *Highest Duty – My Search for What Really Matters*. William Morrow, an imprint of HarperCollins.

Telegraph (2013). Police to think twice about rescuing drowning, www.telegraph.co.uk/news/uknews/1564648/Police-to-think-twice-about-rescuing-drowning.html, accessed 27/09/2013.

Tissington, P. & Flin, R. (2005). Assessing risk in dynamic situations: lessons from Fire Service operations, *Risk Management*, 7(4), pp 43–51.

Trimpop, R.M. (1994). *The Psychology of Risk Taking Behavior*. New Holland.

Truman, H.S. (n.d.). www.brainyquote.com/quotes/quotes/h/harrystru109615.html, accessed 02/10/2013.

Waters, J. (2001). *Police Protection? A Retrospective and Prospective Review of Health and Safety Provisions in the Police Service*. University of Salford.

Weick, M., Hopthrow, T., Abrams, T. & Taylor-Gooby, P.F. (2012). *Cognition: Minding Risks*, http://kar.kent.ac.uk/id/eprint/32472, accessed 10/01/13.

West, R. & Hall, J. (1997). The role of personality and attitudes in traffic accident risk, *Applied Psychology*, 46(3), pp 253–264.

West Yorkshire Fire and Rescue Service Incident Command System (WYICS) (2002). Telephone interview, 27/09/2013.

Wikipedia (2012). http://en.wikiquote.org/wiki/Risk, accessed 20/11/2013.

Zuckerman, M. & Kuhlman, D.M. (2000). Personality and risk-taking: common biosocial factors. *Journal of Personality. Special Issue: Personality Perspectives on Problem Behavior*, 68(6), pp 999–1029.

Bibliography

The titles in our bibliography are informative references (in our view) to assist readers to learn more about providing healthy and safe working environments in the context of organisational development. The authors have accumulated over fifty years' experience in providing advice and opinions on safe systems of work to organisations large and small across the public and private sectors, and the titles noted represent the main titles and standards which we have found useful over the years to underpin our own understanding and professional development.

Together with those titles in the References, they have helped us to understand the place and role for DRA within any type of organisation.

Of course, no list such as this one could be exhaustive, and you may find other titles equally useful.

Books

Anderson, G.M. & Lorber, R.L. (2006). *Safety 24/7: Building an Incident-free Culture*. Intertek Consulting and Training.

Atherton, J. & Gil, F. (2008). *Incidents that Define Process Safety*. Wiley Inter-science.

Barton, T.L. et al. (2002). *Making Enterprise Risk Management Pay Off*. Financial Times/Prentice Hall.

Borge, D. (2001). *The Book of Risk*. Wiley.

CCPS (2007). *Guidelines for Risk-based Process Safety*. Wiley and the Centre for Chemical Process Safety (CCPS).

Cormack, D. (1987). *Team Spirit*. MARC Europe.

Covey, S.R. (1989). *The 7 Habits of Highly Effective People*. Simon & Schuster.

Crainer, S. et al. (1996). *Leaders on Leadership*. The Institute of Management.

Curwin, J. and Slater, R. (1991). *Quantitative Methods for Business Decisions*. Chapman and Hall.

Dalton, A.J.P. (1998). *Safety, Health and Environmental Hazards at the Workplace*. Cassell.

Deming, W.E. (1986). *Out of the Crisis*. MIT Press.

Deming, W.E. (1993) *The New Economics for Industry, Government, Education*. MIT Press.

Drucker, P. (1970). *Drucker on Management.* Management Publications Limited for British Institute of Management.

Eves, D. and Gummer, J. (2005). *Questioning Performance – The Director's Essential Guide to Health, Safety and the Environment.* The Institution of Occupational Safety and Health.

Fuller, C.W. and Vassie, L.H. (2004). *Health and Safety Management – Principles and Best Practice.* Prentice Hall.

Heller, R. (1998). *In Search of European Excellence.* HarperCollins.

Hendy, J. and Ford, M. (2004). *Redgrave, Fife and Machin – Health and Safety.* Butterworth.

Jay, A. (1967). *Management and Machiavelli.* Pelican.

Moss-Kanter, R. (1989). *When Giants Learn to Dance.* Touchstone Simon and Schuster.

Peters, T. (1988). *Thriving on Chaos.* Pan Books.

Tzu, S. (2009). *The Art of War*, trans. by Lionel Giles. First published in 1910. Pax Librorum.

Technical standards

American National Standards Institute

ANSI/AIHA Z10–2012 *American National Standard for Occupational Health and Safety Management Systems.* ISBN 1 931504 64 4.

British Standards Institution (BSI)

BS 8800:2004 *Guide to Health & Safety Management Systems – Guide.*
BS OHSAS 18001:2007 *Occupational Health and Safety Management Systems – Specification.*

International Standards Organisation (ISO)

ISO 31000:2009 *Risk Management – Principles and Guidelines.*

Publicly Available Standards (PAS)

PAS 43:2010 *Safe Working of Vehicle Breakdown and Recovery Operations* (sponsored by SURVIVE (Safe Use of Roadside Verges in Vehicular Emergencies)).
PAS 99:2006 Integrated Management.

United Nations/International Labour Organization (ILO)

International Labour Organization *Guidelines on Occupational Safety and Health Management Systems.* ILO-OSH 2001. ISBN 92 2 111634 4.

Example sector standards

OGP HSEMS 6.36/210 *Guidelines for the Development and Application of Health, Safety and Environmental Management Systems* (oil and gas).
API RP-754 *Measuring Process Safety* (American Petroleum Institute).

Other standards which may be of interest

AS/NZS 4801:2001 *Occupational Health and Safety Management Systems – Specification with Guidance for Use.*
HACCP *Hazard and Critical Control Point* (food hygiene).

Index

Page numbers in *italics* indicate figures, illustrations or tables

Sullenberger, Chesley 'Sully' 38–41, 77–8
Suzy Lamplugh Trust 110
system, use of term 153
system 1 reasoning process 4, 5
system 2 reasoning process 4

tactical decision games and simulators 81
Taylor-Gooby, P.F. 49, 135
Technical Management Board (TMB) 65–6
theory of reasoned action (TRA) 54–5
third party 153
time pressure 64, 103
Tinsley, C.H. 6
Tissington, P. 4
Titus, W. 5
trade unions 51
training 19–20; 3-Level Risk Management Model™ 119; all staff included 129, 131; biases 119–21, 129, 131; and business environment 119; case study 7.1 noise nuisance 121–3; and core values 117; debriefing and feedback 116, 119; decision makers 81; effective 116–17; and expected behaviours 117–18; frequency 116; health and safety 114–15; initial 149; and multi-agency working 116, 118, 130; on-going 151; programme review and modification 117–18; providers 115; refresher 116, 126, 152; and safe

person concept 119; scenarios 118–19; senior management support 115, 118
Truman, Harry S. 119
Turnbull report 88

UNISON 139
United Airlines flight 232 77–8
US Airways flight 1549 38–41
US National Transport Safety Board (NTSB) 77

values and beliefs 4, 5, 6, 7, 54, 55, 56, 93, 120, 153
Vassie, L.H. 26

Weick, M. 49, 135
welly-wanging 14–16
West, R. 7
West, R.F. 4
West Yorkshire Fire and Rescue Service 133
Wiegman, O. 53–4
Wikipedia, definition of risk assessment 26
Worcester Fire Brigade 3
workforce, consultation with 51, defining 153
working at height 17–18
work-related violence 123–9, 137–40
World Health Organisation 123

Zuckerman, M. 7

Milton Keynes UK
Ingram Content Group UK Ltd.
UKHW040057071024
449327UK00019B/613